The Call of Carnivores

Also by Hans Kruuk

Predators and Anti-Predator Behaviour of the Black-Headed Gull. Leiden: Brill, 1964.

The Spotted Hyena, a Study of Predation and Social Behaviour. Chicago and London: University of Chicago Press, 1972 (repr. 2014, Brattleboro, VT: Echo Point Books).

Hyaena. Oxford: Oxford University Press, 1975.

The Social Badger. Oxford: Oxford University Press, 1989.

Wild Otters. Oxford: Oxford University Press, 1995.

Hunter and Hunted: Carnivore Behaviour and Their Relations with People. Cambridge, UK: Cambridge University Press, 2002.

Niko's Nature: A Life of Niko Tinbergen and his Science of Animal Behaviour. Oxford: Oxford University Press, 2003.

Otter Ecology, Behaviour and Conservation. Oxford: Oxford University Press, 2006.

The Call of Carnivores

Travels of a Field Biologist

Hans Kruuk

Drawings by Ineke Kruuk and Diana Brown

PELAGIC PUBLISHING

Published by Pelagic Publishing
PO Box 874
Exeter
EX1 9YH
UK

www.pelagicpublishing.com

The Call of Carnivores: Travels of a Field Biologist

ISBN 978-1-78427-193-0 *Paperback*
ISBN 978-1-78427-182-4 *ePub*
ISBN 978-1-78427-183-1 *PDF*

Cover: Spotted hyena (*Crocuta crocuta*) crouched over kill, taken with remote
camera. Liuwa Plain National Park, Zambia. © Will Burrard-Lucas/naturepl.com

Printed and bound in India by Replika Press Pvt. Ltd.

MIX
Paper from
responsible sources
FSC® C016779

To Saskia, James, Hannah,
Edward and Lyndon

Contents

Acknowledgements

I am immensely grateful to Niko Tinbergen for opening my eyes to animal behaviour, and for the great times we had in the bird hide. My Serengeti and Ngorongoro experience was made possible by John Owen, Director of Tanzania National Parks, to whom I owe immense thanks. Also in Africa I benefited greatly from the wise words and friendship of Hugh Lamprey, and Gus Mills introduced me to the marvels of the Kalahari. Everywhere I worked I was helped by many students, assistants, colleagues and friends, far too many to name them all, but they will know how much I appreciate them, still. And, of course, I gratefully acknowledge the financial support from numerous institutions. The writing of this book would have been quite impossible without the backing, encouragement and criticisms from my children Loeske and her husband Patrick, and Johnny and his wife Alice – my gratitude to them knows no bounds. Thanks also to my sister Ineke, for her touching drawings of the Serengeti hyenas, and to Loeske Kruuk, Lyndon Meir, Gus Mills and the family of Niko Tinbergen for the use of photographs. The Alcock-Brown Trust kindly allowed me to use the splendid sketches of otters by Diana Brown. But especially, I will never have sufficient words for my deep gratitude to Jane, who enjoyed the hyenas, Solomon and everything since them just as much as I did.

CHAPTER 1

In the field

Serengeti house with spotted hyena. (Ineke Kruuk)

THE ENTIRE FLOOR of the Ngorongoro Crater in East Africa, all this huge area of beautiful grassland, lakes and streams, is ours for the first few hours of daylight. There are no tourists yet, and our tent catches the first rays from the sun rising above the forests of the crater rim. Birds sing, zebras bark, wildebeest grunt, some cranes fly over. The two large groups of hyenas in front of Jane and me, some 40 animals, are a striking confirmation of my latest discovery, that these fascinating animals live in large clans which are often at loggerheads with each other. It is another new result of my study here, along with the knowledge that I am watching real hunters here, not the scavengers of popular imagination but hunters, just like wolves, hunters from whom other carnivores such as lions steal. To be able to pursue science in such surroundings is a tremendous privilege, a fortunate and fabulous existence.

I have spent most of my life watching the behaviour of wild animals. How that began, and came to dominate my life, I don't really know. But what I can see now is how a love of natural history turned into science, into observations and experiments that asked how animals' behaviour shapes their existence. Over the years I have written books and scientific papers about my work, about animal behaviour and ecology, and why I believe this to be important. Here, I am writing about the pleasure I derived from being involved in it, about the wonderful natural history, about the rich tapestry behind all my observations in Africa, in Shetland and in many other parts of the world. It is about working with wild animals.

Spotted hyena in Ngorongoro Crater.

Probably, my parents nurtured my passion because of their own enthusiasm for nature in Holland. But seeing how often parental enthusiasm fails to translate into a child's development, I don't think that this is the ultimate explanation for my career. It was something deeper down in my blood, something that even when I was small made me need to be involved with living things, with beings other than people. It made me, when I was five years old, nag my parents to get me a small aquarium to keep sticklebacks and beetles, it made me escape, even at that early age, to go on my own and use my child's fishing net in the moat of our old Dutch town, Middelburg. It made me experiment, as a small boy, to investigate whether cats could swim by dropping a neighbour's moggy into the stream at the end of the garden.

My childhood passion set me off on a path that resulted in me sitting on the bonnet of a Land Rover, watching a pack of hyenas in pursuit of a wildebeest in the Serengeti. It found me hiding behind a rock along the shore of the Shetland islands north of Scotland, timing the dives of an otter just in front of me. Wonderful times. On the way here and in younger days, there were the many butterflies on the buddleia in the garden of my teenage years, the young jackdaw which I collected from the chimney and tamed, the flock of cranes over moorlands in the southern Netherlands, the bright green tiger beetle that I deceived into catching rabbit pellets as if they were flies, the black grouse on their lek, or the field crickets which I caught by tickling them out of their burrows.

The author in Ngorongoro.

Female otter in Scotland.

I was fortunate to be part of a Dutch youth society for natural history, in which we all went on weekend excursions and summer camps. We fanatically pursued our interests in nature, be they birds, plants, sea life, water insects, mammals or whatever. We vigorously addressed adult society, and plugged conservation issues wherever we could. It was only later, at university, that I learned to translate my natural history interests into science, first in plant taxonomy, later in animal behaviour and ecology.

I have little doubt that the greatest scientific influence on me, from my university years onwards, was my mentor and friend, Niko Tinbergen, the Oxford animal behaviourist or 'ethologist', who later received the Nobel Prize for his work. He was a phenomenal naturalist and educator, who used his love of nature, of natural history, as a stepping stone to science. Observe what an animal is doing, ask 'Why?', and that is where our kind of science begins. 'Watching and wondering', he called it.

Niko absorbed me into his research group in Oxford University. There, when watching birds with him, I came to realize that his 'Why?' of animal behaviour can have different meanings, referring to an animal's evolution, or its well-being, or its internal wiring, or its maturation and learning. What 'why' does not refer to is a kind of human reasoning that is sometimes attributed to animals, which Niko considered as 'unscientific'. In his thinking, an animal does not 'want' to do something or go somewhere, but it receives stimuli that result in setting its behaviour patterns in motion, depending on some internal condition.

Niko's bird studies, his *ethology*, treated an animal as a kind of behaviour machine. For him also, right from the beginning, it was essential that behaviour was studied in wild animals, in their natural environment, where the behaviour had evolved. To me, too, this has been a must, a *sine qua non* for my studies, not just after being indoctrinated by Niko but for most of my life and almost instinctively. It is a conviction that was, and is, by no means shared by everybody. But in Oxford, Niko's vision of animals accommodated my interest in natural history, in animal behaviour and in ecology, in the entire relationship between animals and their changing environment. That first

Niko Tinbergen and stick insect in the Serengeti.

natural history involves just watching. Science comes in with asking questions, creating and testing hypotheses and doing experiments in nature, which really is the same as asking and answering questions. That is how Niko put it: he liked to keep things simple and always emphasized how if one could not express science in simple terms, one did not really understand it.

Interest in conservation follows by default. Conservation needs arise from observations, and conservation management needs to be built on the kinds of question that science asks from natural history. Animal numbers, populations, are always somehow limited by their environment, and my involvement has always concentrated on their behaviour as an all-important mechanism: behavioural ecology. I tend to get hit first by the 'wow' factor, by the excitement and astonishment of experiencing a wonderful new environment with fascinating animals, the birds, mammals, fishes, insects or whatever. Then questions follow, with suggestions for answers. They come simultaneously with the strong urge to prevent those treasures being destroyed by people in any way.

I am writing this as I remember my background and interests during marvellous times along British coasts and rivers. I remember the savannahs of the Serengeti, forests of Thailand, an island in Lake Victoria, shores of Shetland, deserts of the Kalahari and Kenya, lava fields of the Galapagos, streams of Australia, seas around Alaska, rivers in the Pantanal of Brazil and many other places. I want to write about these interests because, strangely, natural history and the related science require promotion. There is a desperate need for more biologists to get involved in research on the behaviour and ecology of wild animals, for young scientists to tear themselves away from their computer screens even if just briefly and, essentially, to be with the animals out in the field. Nature now needs understanding, it needs people experiencing it, and it needs attention out there in the wilds, so that we can help it.

CHAPTER 2

Camouflage in an aquarium

Common or Dover Sole.

THE YEAR IS 1959. Looking back on this time, I see myself as a student, with an interest in fish that long predates my later involvement with their predators, with otters. I am studying, working on a project about how fish avoid capture – it is an experience that turns out to be relevant.

Most of my later life will be spent with mammals and birds, but here I am keenly interested in animals such as fish, despite the disadvantage that I can hardly watch them in the wild surrounded by fresh air and wild nature. Fish, for that reason, leave me frustrated. Watching them in an aquarium, as I am doing here now, is interesting, though I cannot help but think of nature outside, think of yesterday, deep winter, when I was skating on a lake here in Holland. Then, whooper swans winged overhead, calling, having flown in from Scandinavia. My heart aches.

During a Dutch urban night I am sitting in front of a huge fish tank, in a dark and dank cellar of a laboratory in Utrecht. I am in my later student years, yet I still feel unsure about what I want to do with animals. I want to study them, to be out in the wild with them – but to do what? For my present student project I am trying to work out some answers to problems around the behaviour of a marine fish. Not just any fish though, but one that is commercially quite important. It is there somewhere in front of me, but right at that moment I cannot even see the animal. The laboratory cellar is a grim place, and I think rather wistfully about 'the great blue yonder'. I daydream of birds and plants, of all the different ducks I saw yesterday when skating, in a patch of open water in the ice.

There must be something more to 'biology', the subject I am reading for my degree, than this cellar and these fish behind glass. They are fascinating, of course, and so is the problem I have been set. But I feel restless, stimulated by my recent reading of a book on animal behaviour in the wild, by Niko Tinbergen. One of his points that really rings a bell with me is the need to get away from my laboratory approach, away from the idea that one can and should control everything around an animal, then study the effects of experimental changes. Niko's writing agrees with my natural inclination to just watch an animal in its natural milieu. After watching, says Niko, after that and as a science experiment, perhaps make small single changes, to see what happens.

In the soft, dark red light that enables me to see in these dungeons what happens in the seawater tank even at night, the sandy bottom of the huge aquarium in front of me suddenly heaves. A sole, a lovely, delicate flatfish, lifts itself from the sand and, with a quick set of movements, performs what I am calling the Omega jump. It is a vertical course through the water best described by the Greek letter Ω, landing again close to where it had been before, but now without a layer of sand on its back.

Before the upheaval, the sole's whereabouts were unknown to me, in one of the best-camouflaged hiding efforts one can imagine. Now, in what to the fish is the dark (because it is insensitive to red light), the sole sacrifices camouflage to its need to eat. Off it shuffles, slowly testing the bottom of its watery abode for the presence of worms, shrimps or sand eels. The Omega jump is the beginning of its nocturnal activity, and also the beginning of its vulnerability to the huge trawl nets that pound the North Sea. Omega is my student project discovery (and I am hugely proud of it).

My little study aims at discovering why soles are mostly caught at night, in the valuable Dover sole fisheries industry that covers the southern North

Camouflage removal: the Omega jump.

Sea. Fishermen want to know how they can improve catch efficiency, though I myself am more interested in the behaviour of the fish, in its striking ability to avoid predation, and its avoidance of being scooped up by mankind. Before, there was just nothing in the sand that indicated the presence of a sole to an uninitiated eye, though with the benefit of experience one can spot the sole's eyes and mouth. The camouflage of this fish, including its adaptive behaviour, is superb.

It makes me ask how it manages this, how exactly its behaviour works to achieve such a result, and if it still has any predators that are not deceived by its perfect camouflage. Exactly the kind of question that, as I learn later, Niko Tinbergen would have asked. There is, of course, currently no way to actually watch what a fish like this does in the wild, at night and deep down under the waves of the North Sea. I need this artificial set-up in the cellar, alas. At least it does enable me to play around with the light, measure it and slowly decrease its intensity at the end of 'daylight', to study what makes the sole perform its camouflage behaviour. I am excited to see the sole's Omega jump usually at about the same light intensity. Then the night-long shuffle starts, the quest for food – and later, when I increase the light again, I watch the fish digging in, with rapid head-and-body beats on the sand, a slow vibration that causes a cloud of sand to descend on its back – and (almost) total disappearance. A predatory fish, such as a cod, would be hard pushed to spot a sole in the sand, and a trawl net dragged over the surface would go clear over the fish without scooping it up.

There still is that nagging doubt that what I am seeing here, in the aquarium in laboratorial dungeons, is not necessarily what happens in the wild. Hence, in the following months I take some week-long trips on the North Sea, on board a Dutch fishing boat. The trawler *Adeodates* fishes mostly on the Brown Bank and the Dogger Bank, in the centre of the North Sea. Every three hours, day and night for a week without a break, the trawl net goes overboard, takes all that lives on the bottom and is hauled up, emptied of its large catch of fish, and immediately lowered again. The five crew sleep for an hour, then work for two hours, then sleep again in their narrow bunks, day in, day out. It is an exhausting existence, but it pays well.

I am tolerated on board by this bunch of friendly ruffians while I pick out my fish, the Dover soles, from the rest of the catch and gut them while they are doing the other fish. I want to see what soles are eating and when. The contempt of these Dutch fishermen for the seasick student knows no bounds, watching me being sick into an old wooden clog in my bunk while they stand there looking, shaking their heads, commiserating and mocking. Somehow over the days at sea I manage to collect my feeding data from all this, collecting the guts from lots of soles caught at different times of day.

Being at sea, fishing like this, is a profound experience. It is the sea, the swell and the waves, the horizon, the gannets and gulls, the experience of being in this huge expanse. The net comes in with its enormous content of different kinds of fish and other creatures. All this makes up for the rocking of my confidence by the fishermen's teasing mockery. And I get the data I wanted. Analysing in the lab at what time of day the soles have food in their stomachs,

Chicks and hatching egg of black-headed gull: dependent on camouflage.

or at different sites lower down in the gut, the presence of their prey, worms and sand eels, shows that soles feed at night. It confirms what I thought all along after watching them in the tank. It is a story that makes my first ever scientific paper, a story of camouflage and foraging, and the related behaviour patterns responding to light intensity.

That simple flatfish makes me think about what stimulates animal behaviour, what are the triggers – in my case, what causes the Omega jump? And at least as important, what are the effects of the behaviour, how does it protect the animal? Obviously also, the Omega jump evolved long before the sole met its most important predator, today's trawler net, so I should look at whether and how natural predators are taken in by it, to understand the biological function of the sole's behaviour repertoire.

It is my first study of animal behaviour, clearly 'applied science'. Yet, applied laboratory work is not the kind of science that I want to commit myself to in the long run. I need to watch animals, of whatever kind but preferably birds and mammals, in their natural environment. Feeling itchy-footed, it is pure luck on my part that, just at the time of finishing my fish project in Holland, I find a place for my needs in Oxford. Niko Tinbergen, my all-time hero of ethology, of animal behaviour studies, has a field study running there on camouflage in gulls. And through friends, I hear that he needs a student assistant.

CHAPTER 3

Gulls and their enemies:
foxes and hedgehogs

Black-headed gull colony.

O N THE SHORES of the Irish Sea, in the north-west of England, is – was – a gigantic colony of many thousands of black-headed gulls. It is – was – the largest in Europe, a colony that has been known in historical records for centuries near the village that the Romans called Glanoventa and is now Ravenglass, with its magnificent beauty of large dunes and sands, and its abundant wildlife along the sea and the wonderful estuary. I walk around there watching the mergansers on the river, the otter tracks along the muddy banks, the natterjack toads in the little dune lakes. I track the foxes across the huge sands, and I marvel at the dense colonies of sandwich terns.

In 1960, I am here for the gulls. In the youthful exuberance of my student days I feel that I know the gull colony, I feel that I know it like a single creature, with all its excitements, its woes and miseries in every detail. In the mornings, I feel the mood of the colony, I feel if there have been dramas in the night, I know of its successes and progress through the season, its accidents and failures, and I think I know who and where the enemies are.

Gulls' eggs, three in a nest, are beautifully camouflaged, with dark brown speckles on the egg-green. The black-headed gulls' enemies such as crows, herring gulls, people, foxes, hedgehogs and others are all after these eggs, often quite successfully so. Seeing their depredations, Niko Tinbergen has long wondered if the camouflage is any good, if it helps the gulls' survival, and if it is therefore worth all the special efforts made by the nesting birds. Camouflage was my interest in my fish project, and here in Ravenglass I am one of Niko's students to whom such of his questions are entrusted, while living in a caravan next to the huge colony. To me, life knows no greater joy.

We are interested especially in one tiny detail of the camouflage protection. When a gull's egg hatches, it leaves the parents not only with a chick but also with an empty eggshell in the nest, a large, hollow and conspicuous object with a white broken rim, possibly a big impediment to nest camouflage. Niko feels that an eggshell must be a major give-away to enemies, especially since any newly hatched chick is an all-time favourite for every predator. Even the other black-headed gull next door is keen on swallowing a juicy chick. So the cavernous empty eggshell could well make nest, eggs and chicks much more conspicuous to an attacking crow, spoiling the camouflage – but does it? Parent gulls fly off with the empty shell and drop it at some distance – is that really worth the effort, especially as parental absence exposes the nest to danger?

I am watching a simple experiment designed by Niko. I have laid out some black-headed gulls' eggs in a dune valley away from the colony, half with an empty eggshell next to each of them, the other half just alone as single eggs. Hours pass in my bird hide high up on a dune slope while I watch this set-up, and I produce heavy science with my watch, pencil and notebook. Crows fly past, a couple of large herring gulls circle overhead, but nothing happens to the eggs. I entertain myself by watching a couple of shelducks visiting rabbit holes, prospecting for a suitable nest site. Across the valley, a single crow alights on a small dune, sits and waits.

After an age, the crow wings across, flying straight at me in the hide some eight or ten metres above ground, before suddenly stalling, and dropping down next to one of my experimental eggs in a marked place. From my distance it is difficult to make out if this is an egg with or without an empty shell, but whatever, the crow flies off with its prize, out of sight. Soon I see the same thing again when the crow returns, a quarter of an hour later. This time, a flying-over herring gull also notices what the crow is up to, and it does something similar. Results now come in fast and furious, and I am pencilling away inside my small canvas shelter, sitting on an old milk box.

A couple of hours later, and when I guess that something like half the eggs must have been taken, I call a halt, break out of my hiding place and collect and record the eggs that are left. More than three-quarters of the eggs with an accompanying eggshell have disappeared, but only one-quarter of the eggs that had not been flagged by an empty shell. It shows dramatically how that single empty eggshell endangers the remaining eggs in the nest, and shows the potential effect of that one, very brief behaviour pattern of the gulls, their 'eggshell carrying'. In other experiments with eggshell models we discover what it is about an empty eggshell that induces a gull to carry it away. With these different approaches, Niko teaches us how to distinguish cause and effect in behaviour, one of his major contributions to animal behaviour science.

Niko delights in such simple experimenting with details of behaviour and camouflage, in finding out how the black-headed gull organizes all these details, and how that affects its survival. For me, it opens a new way of life, a new science. Niko and I click: for me he is the ultimate naturalist. We roam the dunes together, me the young and eager student with him teaching me how to think, how to watch birds, how to experiment and how to track foxes or hedgehogs across the sands.

Following those first years as Niko's student assistant, I am now doing my own PhD project here in the Ravenglass dunes. I almost feel as if it is me who is pulling the strings of all the different predators around, of the crows that steal the eggs, of the foxes and badgers that murder the adults on the nests, of the hedgehogs that slowly devour a living chick, of the peregrine that swoops across after a single bird. It is all part of my big question: there is this huge glut of many thousands of nests here with eggs and chicks, and there are lots of predators – how does this all work out together? I can see the tremendous response from the gulls to their enemies, their magnificent anti-predator organization with all its beautiful corollaries in behaviour, colours and choice of surroundings.

I realize that I am only an upstart PhD student, and while beginning to think that I know how and why the gulls do what they do, in reality I am still only at the beginning of my predator project, just looking for answers and understanding. And while I look for insights in the behaviour of the gulls, I also want to know how the predators do things, how they get the better of the black-headed gulls, and I try to match predator and anti-predator behaviour.

Sitting in a small bird hide, I am conducting an experiment that should show me what prevents the huge masses of black-headed gulls being totally annihilated by their enemies. I may sometimes be feeling that I know some of the answers, but I only have a few ideas about how it all works. My experiment should suggest whether some of these ideas are right. There are times when I have a hypothesis that runs away with itself and I may forget that I still have to test it. Such confidence is dangerous.

This morning I am sharing the smelly and dirty canvas bird hide with Niko himself, who is trying to imprint such wisdom and need for modesty on his impatient student. He sits next to me on his usual plastic fish box, smoking like a chimney, keen and excited like a student himself, as we watch the predators and the gulls. He inspires me with the importance of 'just watching', watching what happens in nature, then probing for answers with experiments, if at all possible.

Niko to us, his students, is known as 'maestro'. On his fish box in the hide he is as relaxed as he ever will be. He clearly loves being with his people out in the field, and here in the hide he is wonderfully alive, very alert in his observations of the birds, delighted if he sees something before I do, pontificating about them and about science. He picks on mistakes and weaknesses in my experiments and my notes, often to my great annoyance – and then suddenly, fag in hand, he comes out with such nonsense as 'How much wood would a wood chuck chuck if a wood chuck would chuck wood?' A wonderful man, as I realize once again when many, many years later I write and publish his biography.

The experiment I am watching now is about predation by crows and herring gulls on broods of the much smaller black-headed gulls, and the protective behaviour of the latter. The black-heads vigorously attack the predators when they get too close to the colony, yet many an egg and small chick gets stolen. So, do these attacks on the predators protect the nests at all, or is that something that we are just assuming? I set out some hens' eggs both inside and outside the colony area and watch what crows and herring gulls do to them, and how successful they are in their predation. Black-headed gulls are not interested in hens' eggs.

I feel that this is the right way to do this, when I leave nature more or less undisturbed, and change as neatly as is possible just one simple variable (in this case predator food close to or far from a gull's protection), then watch what happens. I am in my element, watching wild animals while playing small tricks on them, to see how their organization works.

The hens' eggs experiments demonstrate how efficient the black-headed defence system is. Without the protection of a colony, the eggs I put out are taken by the predators in no time at all, and in quantity. The crows bury and hide eggs they cannot immediately consume, so they may take dozens in a very short time. After all my experimental eggs outside the colony have gone, the crows brave the attacks of the black-heads and take a few prey from inside the colony, getting a great deal of stick from the black-heads. The larger herring gulls are completely put off by such attacks, and they only take unprotected eggs. There are successes on both sides, of defenders and of attackers.

Niko and I spend a long morning with all this, pointing things out to each other and discussing the various happenings, before heading back to camp and looking for animal tracks in the sands along the way. Endless cups of coffee wait in the caravan, as well as two other PhD students in their tents next to it. Altogether, our Ravenglass camp is a rather spartan arrangement, but it does the job – it is a happy place.

The camp is exposed: there are no trees, just dunes, sands and grass. During one year I need to be there even before the gulls arrive in early spring, to see what happens on their mass arrival and how the predators react. I am reminded of one ghastly night in February, during a howling gale with rain, when the tent is shaking and cold and I huddle in my sleeping bag, feeling miles away from anywhere and anyone. In the pitch dark I suddenly wake up with somebody or something soundlessly feeling their way up from the bottom end of my sleeping bag, and I pick up a smell like that of an unwashed body. Torchlight reveals a large white ferret, an obviously tame escapee from somewhere, an animal wanting nothing more than a bit of comfort here in the dunes (which it did not get from me).

The large dune peninsula is my territory, I know every bit of it, I love it. Apart from my actual PhD project on predation and the gulls' defence, there is masses

Walking the colony with my tame hedgehog. (Niko Tinbergen)

of rich background tapestry. There are the migrant birds coming through and feeding on the estuary, the shelducks, oystercatchers and mergansers breeding in the dunes, the calling redshanks, the ringed plovers and lapwings, the skylarks, the ravens and the foxes, and there are the ever-changing wind-blown sands. I learn to read the sands' daily records of who did what where, the tracks of hedgehog, badger and fox, which is my regular early-morning reading matter.

My life of watching and recording predation on the gulls, and recording the behaviour of the gulls when they are challenged, is interspersed by experiments. I move around with hens' and gulls' eggs, watching reactions to them by predators from hides or from distant dunes. I put out mounted specimens of fox, stoat, crow and hedgehog near the nests, and pull the stuffed fox on a sledge along the beach just in front of the colonies, using the Land Rover. I watch the reactions of the gulls to the different kinds of predation, and the gulls' responses to my predator models.

A late night in May. It is quite dark, and I am walking along the beach, just outside the bird colonies, surrounded by the eternal sounds of the sea some distance away across the wide sands, and by the quiet gurgling and calling from the gull colony. Despite the quiet I am all alert, hoping to see a fox on the beach, or whatever.

From the vast colony grounds comes the sound of a gull screaming loudly, just one single bird. It is a horrible noise, it goes on and on from the same spot, without change. Climbing the dune and following the light of my torch in between the nests, I arrive at a one with a large gull chick, almost adult size, with a hedgehog on top of it. The hedgehog is tearing away at the bird's rear end, eating it alive, from the preening gland above the tail onwards and into the body cavity. The chick is looking over its shoulder, unable to move, and screaming incessantly. The horror will not leave me, especially after finding out later that this is a quite common occurrence. To me, a hedgehog will never again be that friendly, sweet creature of our postcards, but a significant predator on birds, with a nasty streak to it.

Following this, a captive hedgehog does experimental service within a low fence inside the main gull colony. The fence allows the gulls to come and go from their nests but keeps the hedgehog inside the enclosure. The nests contain only eggs with the incubating adult gulls, and by carefully moving the nests over small distances I have created a densely crowded area inside the enclosure, as well as a sparse one. Watching from a hide, I see that the gulls accept my manipulation without problem, sitting on the nests and viciously diving or pecking whenever the hedgehog approaches. The predator carefully avoids the densely crowded area, and takes eggs only from the sparsely populated part – demonstrating a clear advantage for the gulls of nesting closely together, and the usefulness of co-operating against the prickly enemy.

In the Ravenglass camp I also have a tame crow that I reared as a nestling, called Caracho. He preens my ears and face, and rubs his head against mine. He joins me on walks, flies from my shoulder and comes when called, wonderfully useful when I am studying the reactions of the gulls to a crow, their predator. Interestingly, Caracho has great respect for the gulls when they dive-bomb us, and he clearly hates me going close to the nests.

During my many months in the Ravenglass dunes I usually think of little else but my experiments, my observations of the predators, of the harassed black-headed gulls and the rapacious herring gulls, the many gulls killed by foxes. I am obsessed by hedgehogs and crows, peregrines and harriers, and I try to

Black-headed gulls killed by a fox, next to their nests.

Night-time killer in the colony: fox.

understand how to analyse the data. The camp is visited by other Oxford students, some for several months but often they come for only a few days, and we endlessly discuss my results, as well as their own research. My peers – Mike, Ian, Bob and others – are stinging and relentless in their criticisms of my attempts and conclusions, and they are a tremendous help in getting things right. Often there is Niko, too, who is much gentler but at least as devastating in his comments. Together, they cause me sleepless nights, and I have nightmares about statistics.

Yet, all this does not take away from the overwhelming memories of those wild Ravenglass years, of animal behaviour laid bare in the field. It is not only the beauty of the dunes and the richness of the fauna, but the complicated ecology of the bird colony, the way the birds' behaviour fits in to make that ecology work, and how they are adapted. Niko teaches me to recognize the many elements of animal behaviour in wild animals, the fear, aggression, curiosity, sex and others, and how these elements contribute to survival in the face of potential ecological disaster. I arrive at a picture of how the birds defend themselves against predators, and what the costs are of the gulls' efforts.

All predators are different. A peregrine is a different danger to a black-headed gull from a hen-harrier, a crow represents another hazard again which is also different from the danger from a herring gull. A fox is a major menace, but its danger is very different from that presented by stoats, hedgehogs or people. The gulls are adapted to each of these variations as well as they can under the circumstances, they recognize all these predators for what they are and what they do, and they respond accordingly with different mixtures of aggression, fear, curiosity, attraction to each other and so on. They also have to balance their needs of defence against other requirements, such as their need to forage and to defend a territory against neighbours. This is what my PhD thesis is about, later published as a book.

All this happens well before the days of 'behavioural ecology', the big fashion in biological science that will come later still. Ravenglass is one of its smaller cradles, one of the pillars under Niko's Nobel Prize. Some ten years after those days spent in dirty bird hides, I organize a large international conference on behavioural ecology in Cambridge, the first ever such meeting in Britain.

When I am writing this decades afterwards, sadly, the magnificent dunes, the shifting sands and the gulls of Ravenglass are no longer. They have been removed from the face of the earth in the space of just a few years, by the dirty broom of pollution. Marram grass, ferns and brambles have taken over from centuries of sands. The gulls never returned, a drama that is made even worse by the realization that this huge colony had been there for hundreds of years, at least since the sixteenth century. The probable cause of the disaster is aerial nitrification originating in industrial Britain, changing the vegetation. With the gulls, most of the other prominent fauna also disappeared, such as the sandwich terns, the little terns, the oystercatchers and many others. I have tears in my eyes when I see what has happened. The estuaries are still there, and the beach – but the dunes are a sight that is heartbreaking to anyone who knew the area before. Conservation has come too late.

My Ravenglass years and the experience of working with Niko leave me with the certainty that I took the right turn when deciding on a life with wild animals. Ravenglass gives me the insight that there is a huge and marvellous wild world out there, full of problems of nature, animal behaviour, ecology and conservation, and I know that I can contribute. It leaves me itching to travel and get out further.

CHAPTER 4

Serengeti: hyenas, lions and the dusty track to Seronera

Spotted hyenas in the Serengeti.

IT IS 1964. Still writing up my PhD thesis in Oxford, I happen to attend a lecture by a visiting scientist, as one does at university. John Owen, Director of Tanganyika (later Tanzania) National Parks, talks about the enormous Serengeti, a place little known to us students. He shows a film, *Screngeti Shall Not Die*, made by the German conservationist Bernhard Grzimek, about the quite unbelievable diversity of animals. John describes the horrible threat hanging over the Serengeti, of a government wanting to cull vast masses of wildebeest and zebra in order to feed the starving peoples in Africa.

John is a compelling lecturer, a large man with a chubby, friendly smiling face, rarely without his pipe. His audience is rapt; the thought of all these animals being slaughtered is abhorrent, the environmental consequences of such an action totally unknown. There is hardly any science involved in the governmental decision, hardly any ecology being conducted. What would happen to the Serengeti, to the enormous migrations there, to all the lions, leopards and cheetah? How would a mass cull affect tourism, with all its benefits to the country? John leaves us with the thought that Africa desperately needs scientists – ecologists and animal behaviourists. No place more so than the Serengeti.

After the lecture, I go and have a word with John. We get on, and he tells me he is on his way to visit Prince Bernard in the Netherlands and some Dutch government officials, to talk about money for the national parks. Three weeks later I have a job, to research the Serengeti carnivores and what their fate would be after the planned mass cull there of wildebeest and zebra. Half a year later my wife Jane and I are on our way to Africa.

Only months after that I am reminiscing, sitting on top of one of the largest rocks near our house in the Serengeti, with a pair of binoculars. I watch the world go past. The open, grasslands plains with the Masai name 'Serengeti' have wonderful horizons. Here and there, stacks of huge, smoothly rounded boulders, taller than the tallest tree, break the contours. They are home to curious nooks and crannies of vegetation, with candelabra trees and the odd flat-topped acacia. These *kopjes* are fabulous for their views, and in the vastness of the open grassland plains they attract animals. Little antelopes called dik-diks make the *kopjes* their own; sometimes I see a leopard on a rock, and a pride of lions may be sprawled over the granite.

I think back on my own arrival here with Jane. I think back on the fortunate course of events that brought me into Africa. Good memories and, all the while, I am scanning the grassland plains from my vantage point. My binoculars keep me involved with a family of zebra, close to them the small 'tommies' (Thomson's gazelle), a few Grant's gazelle and a couple of other antelopes, the larger *kongoni* or hartebeest, and on top of a termite hill a single topi bull. A golden jackal walks past with some delicacy in its mouth. Far, far back is a huge herd of black wildebeest; I can hear their distant grunts. The richness is baffling, all is quiet, and my thoughts drift away again, to our first arrival.

I still see that endless, dusty, gravelled track from the town of Arusha into the Serengeti, as it winds its way down from the rim of the Ngorongoro Crater, down

Wildebeest among Serengeti acacia (fever) trees.

Wildebeest migration across the Serengeti plains. (Ineke Kruuk)

from the clouds over forested heights, down from the trees with their curtains of lichens hanging almost across the track. Once out of the cloud forests, dust takes over, and the spectacular views over the plains of the high plateau make it difficult to keep my eyes on the treacherous dirt road, while driving my new Land Rover.

The first sights of wild African animals stay deeply engrained on my mind. We are staggered by those first giraffes and elegant impala, just outside the outskirts of Arusha, and the elephants and buffalo on the road along the rim of the Ngorongoro Crater. Then the Serengeti itself, across the famous Olduvai Gorge, with wildebeest and zebra, endless herds of them.

A spotted hyena, my first, lies in a small muddy puddle next to the track, on our way to the Serengeti headquarters in Seronera. A good-looking female, she wallows in the cool mud, a friendly look on her face. Jane and I cheer it. I think it is then and there that I decide to really concentrate my study on that animal, on that species, in favour of all the other carnivores that I am still expecting. The Serengeti has more than 25 species of carnivores, and John Owen initially suggests that I study all of them. But for me, studying every single one would have meant drowning in diversity. Even before meeting the famed Serengeti lions and other cats, I decide to concentrate on the underdog, on the much-maligned spotted hyena, which is also the most numerous large carnivorous animal here. On top of my beautiful rock, still immersed in thoughts of how I got where I am and where I want to go, I take some deep breaths.

What needs doing is to find out what effect hyena numbers have on other wildlife, on the vast herds of wildebeest, the magnificent lions and all the others. It needs a study of numbers and consequences, of ecology, which also implies unravelling the totally unknown behaviour of hyenas and of their social life. I must sort out what makes them tick. Niko Tinbergen's teaching at Oxford is deeply engrained in me.

Thinking about how to set about all this, what worries me is the general response I get from the old guard here in the Serengeti, from the park wardens,

The Serengeti, with Thomson's gazelle and wildebeest.

from visiting naturalists, from the many old Africa hands, the 'white hunters'. Comments are along the lines of 'You green academics, what do you want to tell us here? All this stuff about hyenas, we know it.' Fortunately, director John Owen is not like that: he realizes that we need hard facts, rather than tales. Yet something still rankles. These old hands do know a lot, and they shrug their shoulders.

Against their scepticism I get the advice from Niko never to take anything for granted, but to assume that we do *not* know. And John Owen also realizes that we need to establish exactly what is happening with the hyenas, and what they are doing. In the meantime, John has success in persuading the Tanganyikan authorities to drop their disastrous plans to cull the ungulates in the Serengeti, which was the original motive for my study here. John holds that whatever the government says, we do need to know what controls and regulates animal numbers in the national park, and hyenas are an essential part of that quest. Other scientists will study other animals at the same time, such as the wildebeest, zebra, gazelle and lion.

Of course the old Africa hands, friendly and interesting as they are with all their experience and their knowledge of the Serengeti, are right in their assessment of me being green. There are times when I am made very aware of that, some very embarrassing.

During my first few Serengeti weeks I am showing some of our visitors something of our life here. This time it is an eminent Dutch ecology professor, Karel Voous, and his wife. He has just flown into the park headquarters at Seronera and they are staying with us in our house; he is very keen to see Africa, she is rather terrified of it all. After a light lunch, I take them for a drive in my Land Rover pickup, and the animals are putting on a wonderful show.

Bumping along, cross-country, they cannot help being astounded by the size of the landscape around us, the gently sloping grasslands with tall, yellow acacias along the streams. A troupe of vervet monkeys, a couple of warthogs and a few topi antelopes, some impala in the bushes, the masses of different birds, it all makes an overwhelming display for these people who have just landed from the urban crowds in Holland. A leopard is eating a gazelle high up in a tree. I drive away from the trees out onto the open grasslands, with nothing but horizon in front of us.

While gathering a bit of speed, I suddenly notice from the corner of my eye a large, black cobra, curled up and sunning itself on a low termite hill. We are past it by the time I mention it to my guests, who did not see it, and I turn back to show them. Slowly I drift the car towards the shining black snake. I guess it is about 2½ metres long, facing us, and absolutely still.

Cobras are shy but they can be quite aggressive, and they are fast as lightning. When we are still about ten metres away, the strikingly black snake suddenly decides that enough is enough and slides towards what it sees as the nearest cover on this open grassland – our car. I say slide, but it goes like a flash, and in a couple of seconds it reaches the protective shade of my Land Rover, right there in front of us. The snake slides out of sight below us, so I reverse as fast as I can and hope to see it. A scream from one of the passengers.

Reverse, reverse – but not a sign of the snake any more. There is no other cover so clearly it must have got into the vehicle, but fortunately it can only be under the chassis or in the engine compartment. After driving away from the site, with my company in terrified silence, I explain that it could not possibly get into the cabin of the Land Rover itself, and I stop to check.

Leaving my friends inside with the car door firmly shut, I carefully lift the bonnet. The old Land Rovers put their heavy spare wheel in front, on top of the bonnet, so it needs lifting with both hands and I have to put a bit of force into it. Hardly is the bonnet up when I drop it again with a terrific bang and a yell

Serengeti leopard with gazelle.

– the big spitting cobra has attached itself to the underside of the bonnet, its head within inches in front of my face. I lose a lot of my credibility.

Later, at home, I spot the snake deep inside the engine, curled up on the gear box, impossible to get at. Making the best of a bad job I drive the vehicle quite a long way out into the bush and leave it there. Next day, the snake has vanished. The story spreads through the village, my Land Rover spreads terror for days after, the people at the petrol pump shy away from it and refuse to serve it. I realize that I was stupid, I should not have let this happen. I am green here.

Only days later I decide I need to stay out at night to see the hyenas, and with my Ravenglass experience I decide to make a hide. I build a large tree platform in a magnificent sausage tree, a *Kigelia*, at the edge of the plains. It is a lot of work resulting in a splendid hideaway, high up and surrounded by huge sausage fruits. It is a shelter where I can be left overnight to watch the animals and be collected again in the morning. I choose the site because of the many bright white hyena droppings underneath, being clearly an important place for hyenas – a latrine or scent-marking area.

That very evening, when Jane takes me there in the Land Rover for my watch, we arrive to find a herd of elephants. They have just pushed the sausage tree over, my platform is totally destroyed and two elephants are contentedly browsing from the branches. Serengeti teaches me another lesson.

I realize how little I know about the country, or even about the animals I am about to study. And whatever questions I want to ask, all the old Africa hands seem to have seen it all already: 'Everybody knows ...', and everyone refers to the hyenas' reputation of simple, cowardly and ugly scavengers, not likely to make any impact here, not attractive to tourists and therefore uninteresting to the park authorities. Nevertheless, I need to go back to basics, assuming that I have to start right from the beginning with these maligned hyenas, and with Niko's dictum sounding in my ears: 'Question everything, watch and see what happens, then make sure you ask the right questions.'

Weeks later, I am still out one early morning after a long night, when life awakens on the plains. From my Land Rover and only a mile or so from our house, I am watching a couple of 'zebra buses' from a short distance away, the

Camping on the Serengeti plains, near Gol.

Lion courtship in Ngorongoro Crater.

ubiquitous small, striped vehicles that are the workhorses of the tourist trade. They are a daily routine, with local guides taking hordes of visitors to see the spectacular wildlife around, with the morning yet young. The zebra buses are stationary, and several pink faces are popping out of their roof hatches, people clutching cameras. Next to the vehicles are the remains of a wildebeest, with two beautiful, black-maned lions resting close to it. It is a standard Serengeti picture.

About 100 metres further, not far from my car, are a dozen spotted hyenas, ostensibly fast asleep, their mouths bloodied, obviously waiting for leftovers from the lions' meal. They are of no interest to the visitors. 'Get a picture of the scavengers!' cries someone from one roof hatch to the other. It is an oft-repeated scene here. The comment is self-evident.

Except, except that I happen to know what has happened before and what is happening now; I know that these guides and their tourists have things diametrically wrong. Because I have been watching these same animals for over four hours now. I have seen the excitements and dramatic events that completely overturn our ideas of hyenas.

In the dark of night two of them, two of the hyenas now prostrate here near the car, walk the grass plains. One starts a chase, the other joins in and together they bring down a wildebeest bull. Many more hyenas join. They squabble and make a terrific noise of yells, growls and screams, audible for miles. They have eaten only little when the king of beasts turns up: two magnificent lions appearing from the dark, running into the hyena mêlée, grunting deeply.

The hyenas utter their soft, staccato alarm call when the lions first appear, then they scatter in all directions, away from the running lions. The two big cats have a sniff around, settle on the generous leftovers of the carcase and eat, tearing away at the soft haunches of the wildebeest. The hyenas withdraw to a safe distance and stay there, eventually nodding off to sleep but still well aware of what is happening. The lions eat their full, and fall asleep next to the kill. They are still there for hours afterwards.

Hours later the sun rises, birds begin to call, high in the sky a vulture appears. In daylight the zebra buses arrive, the tourist guides attracted from

Lion chasing hyenas from their kill, Ngorongoro.

afar by descending vultures. Lions eat, hyenas sleep. And from the tourist buses come the inevitable comments about predator and scavenger. Much later the two magnificent lions will walk away, their big black manes fluffed up, cameras click and hyenas finish off the remains of the carcase. Sure, the hyenas do eat remnants, but these are leftovers from the hyenas' own hunting effort. They are decidedly not scavenging, it is the lions who do that.

This is the first time that I see what really happens out there in the dark between lions and hyenas. Later I find that everywhere here in the Serengeti, but especially in the nearby Ngorongoro Crater, lions steal prey from hyenas. It happens at night and, when people find the kill in daytime, the lions are still there, with the hyenas waiting in the background. In following years my colleague George Schaller establishes that, in fact, on the Serengeti plains lions do kill more often than they scavenge, especially the lionesses. But the above scenario, the theft of the hyenas' prey by lions, is common and it is one of the reasons for the hyena's unsavoury reputation.

At this, my first encounter of the truth behind the misunderstanding, I myself am flabbergasted and excited. It is one of those lovely findings of something new, of something totally unexpected from a common animal. Niko is right, in order to establish how nature works we need to observe, not make assumptions, and be very careful with conclusions.

Of course, I still want to know why the hyena's life is so different from what is commonly assumed. How do these animals manage to catch such large prey? What kind of an impact do hyenas have on populations of all the grass-eaters? What is their social life like: is it just chaos, or do they go around in closely knitted packs, like the wild dogs here, or wolves in more northern latitudes, or do hyenas join up in 'prides' as the lions do, all co-operating in hunting? Apparently it is none of these, and I need to go back again to Niko's advice.

One way to get insight from an unusual animal is to live with it. It helps understanding, and it is a profound experience. That is where Solomon comes into the story.

CHAPTER 5

Solomon, the hyena in my bath

Game-watching from the Land Rover.

THERE IS SOMETHING PREPOSTEROUS about me driving cross-country towards the horizon of the Serengeti plains, covering miles and miles, with a tame hyena on the seat next to me. We are both sitting upright, looking over the dashboard, intently surveying the African grasslands around us in the sturdy, two-seater Land Rover. Solomon, a friendly, soft-haired animal the size of an Alsatian dog, is looking for anything that moves, and I am on the alert for anything that could stop us moving, such as a warthog hole. It is early in the day, and I could have been quietly eating my breakfast on the veranda at home, with Solomon doing one of those things that hyenas do, such as sleeping in a muddy hole, or giggling with his mates. Fate and science bring us together here, and it looks as if Solomon is enjoying it every bit as much as I do.

A pride of lions is stretched out in the shade of a large, solitary, flat-topped acacia. I can pick them out from a long distance, as I notice their ears showing above the grass. I approach slowly, in second gear, careful not to drive straight at them so as not to frighten the animals. But there is no need to worry: the lions hardly take any notice. There are six, three lionesses and three youngsters. I am very keen to see how Solomon reacts to them because he probably has never seen a lion in his life. In the wild, lions and hyenas are the greatest of enemies, stealing each other's food, and they kill each other if given half a chance. Even from our house in Seronera, in the middle of the Serengeti, I often hear the most ferocious and noisy battles between them.

In the car I get within ten metres of the snoozing lions, and Solomon does not show the slightest concern. A day or so earlier I had taken our Tanzanian cook, Lohai, in the Land Rover, sitting where Solomon sits now, and when I drove close to a lion pride, Lohai's response had been very different. He leaned away from the closed car window, away from the lions, his head close to my shoulder, his face a terrified smile. Not so Solomon. Upright, the hyena looks at the big cats, and that is all. He hardly turns his head. It seems he just does not recognize them for what they are.

Until, that is, I drive away. After I start up again, slowly increasing our distance from the lions, circling around and away from them, I unwittingly come downwind of the pride, as I notice from the drifting dust. The lions do not stir. Solomon yelps and quite literally hits the roof of the Land Rover, apparently beside himself with fright. He is totally terrified, as one would expect from a wild hyena encountering a lion.

Driving away across the open grassland space I contemplate this, to me a fascinating observation. The mere sight of lions obviously does not do much to a hyena. But the fear of their smell, something that cannot come from Solomon's experience, is one of the things that keeps hyenas out of harm's way, an arrangement that makes sense for animals that are active mostly in the dark of night, as hyenas are.

I remember then that there is something else that I had noticed earlier about the behaviour of wild hyenas, and in fact of almost all carnivores I have watched. When these animals approach either another one of the same kind, another carnivore or something new or worrying (such as when hyenas approach lions), they almost always first circle around the target until they are downwind. Only then do they come closer, all the time sniffing the air. It fits neatly with Solomon's fright in the Land Rover. Later that morning, I drive home to our small wooden house in Seronera, pleased with myself and with Solomon.

The Serengeti around me here is heart-rendingly beautiful. It is a beauty so enormous that small observations, such as this one with Solomon and the lions, threaten to sink into insignificance. One could doubt the importance of the fear of a single hyena in the scheme of things, when seen in this landscape of horizons, of flowing hills in the distance, of tall, yellow acacia trees along small rivers, of endless herds of wildebeest. What priority a single, simple predator, when on my way back home across the plains I drive past the exquisite Lake Magadi, lined with flamingos, eyed by a few large, black buffaloes?

In my own, somewhat blinkered way, I am trying to contribute to the preservation of all I see around me here. I am a researcher. I am trying to understand and trying to make others understand how all this works, how its ecology functions. I need to know how predators and prey, scavengers and grazers work together to keep the system going, and how we, with the management of the national park, should or should not interfere to keep all this for perpetuity. Solomon is one of my supports in the understanding of the most numerous large predator in the Serengeti. A still largely unknown character, feared, detested, maligned but absolutely fascinating, with a social life that I now

Solomon gets a snack.

know as somewhat reminiscent of that of people: a life in clans, with female superiority. The species may still be enigmatic, but I have no doubt that the hyenas' presence in the Serengeti puts a large, important stamp on much of the wildlife, and therefore on the entire ecology of the area.

Solomon is a wonderful and quite cuddlesome animal, and I have become closely attached to him. Initially, he joins us to help with science, with my study of wild hyenas, because I want to get under the skin of a hyena, to get the feel of so strange and unusual a creature. But then, almost before I realize it, Solomon becomes a story in his own right, and a very lively one.

Soon after Jane and I move into our Serengeti home in Seronera, one of the park wardens, Myles Turner, calls in to say that his ranger Kimali had seen a very small, pitch-black hyena cub in front of a den, in the western part of the park. Myles knows that I want to keep a young hyena, and he is interested. Jane and I join the rangers to collect it, some 40 kilometres away. It takes a bit of digging, and for Kimali to stick his head and shoulders in and reach deep down into the den, and out comes the sturdy, solid little animal that just fits into my two hands, with short, soft pitch-black fur, eyes wide open, and sharp nails and teeth. He must be less than two weeks old.

Once home, Solomon has a ferocious appetite for milk, and whenever not totally saturated, his high-pitched, trembling screams ('miiiiiiiiiilk') can only be satisfied by the bottle. He tolerates cuddling, but his sharp milk teeth are constantly working away on anybody or anything, and even at a very early age he has what in people we would call a strong personality. Now, from research many years later, we know that between hyena cubs, which are normally born in litters of two, there is terrific competition at a very early age, with ferocious fighting even between cubs that are only a few weeks or months old.

Jane in tug of war with Solomon.

Often this results in one of the cubs being killed by its own sibling, even before they emerge above ground from the den. It explains a lot about the early days of Solomon, about his aggression as a small cub. In later years I sometimes feel guilty about having taken a wild cub to keep at home with me but, realizing the fate of cubs that are left alone, I think that perhaps for the cubs it is a lifesaver that I did so.

Soon, Solomon with his warm character becomes much more than a study object, and he is now a personality in the house, a companion. From the house, he goes out on expeditions on his own into the Serengeti wilderness. He is active especially at night and, just as with a domestic cat, more often than not I have no idea what he is up to. He is, in fact, more cat- than doglike in his character: not surprising, as the hyena family is more closely related to felines than canines.

Our house is a wooden, prefabricated bungalow, standing on stilts about one metre above the surrounding savannah. The landscape that we see from our veranda is an endless wild grassland, with flat-topped acacia trees, and we face a distant high, rocky hill called Nyaraswiga. The 'village', Seronera, is a group of scattered houses near the park headquarters, the centre of the national park. From our windows we are not aware of other houses, and there is no fence around a garden, or around the village – we are part of the Serengeti, and the animals are right there with us.

We may wake up at night as our house, often deprecatingly referred to by us as the 'hen-house', shakes when a buffalo rubs its flank against a corner, or when a waterbuck eats from our planted basil, causing a pungent smell. A genet cat calls in and enters, and we hand-feed it. On one occasion I wake up to the sound of pouring water and, jumping out of bed to check on taps or burst pipes, I find a giraffe blacking out the view and relieving himself outside the bedroom window. Masses of wildebeest and zebra often thunder past, a topi antelope stands on an anthill, every so often there is an elephant in the distance. The night brings unforgettable sounds: roars of lions, many calls of spotted hyenas, and of owls.

Solomon often sleeps under the floor outside behind the stilts, with other denizens such as hyraxes or the odd snake. He wakes up for a game when someone walks past, scaring whoever appeared to be asking for his attention.

When he does stay with us inside the house he can be quite destructive. Many are the books he chews, including Jane's favourite cookery book, and I catch him playing catch-the-record when I have the *Carmina Burana* on the old battery-operated record player. There are endless tugs of war with pieces of clothing or towels or carpets, the dirty-clothes basket gets emptied, as are the contents of a cupboard where he goes to sleep.

Sometimes, when Solomon goes missing, I go out and try to find him. I notice that one place Solomon really likes in daytime is a group of boulders with small caves, a *kopje* about half a mile from our house. As in other such *kopjes* in the Serengeti, there often are many hyena droppings between the stones and bushes there, and Solomon likes to be fast asleep somewhere in a crevasse. Because of him, I initially spend a great deal of time visiting this and many other such places across the Serengeti plains, just to find out what hyenas are doing there. Well, they sleep.

Interestingly, I also find that mankind had used the *kopjes* extensively in days long gone, people who lived in these areas many years before it was a national park. There are neolithic remains, obsidian flints, bits of pottery, and there are cup-marks on flat rocks where people played the game of *bao*, a game that is still popular to this date but now on wooden boards. What stirs my curiosity most is rock drawings and paintings in shelters under the granite boulders, often obviously recent. Solomon, however, appears more interested in the shady clefts between the rocks, and usually I need some food to wheedle him out of there, back into my Land Rover.

I am lying in a hot bath, after a long day of heat and dust, with the night-time sound of cicadas coming in through the window, and a distant wildebeest grunt. I soak it up, contemplate my toes, think about my research, and doze. Then, the wooden house vibrates on its short stilts under heavy footsteps, and the handle of the bathroom door is turned. The door bangs open and Solomon bursts in, squealing with delight.

With little hesitation he jumps into the bath, first on top, then next to me, rolling in the water. It is the end of my meditation, as a large, Alsatian-size hyena does not leave an awful lot of space in a bath, and my vulnerability needs

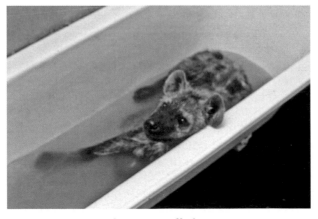

Solomon expelled me.

protecting. It is the first of several such occasions, and I love it when he comes in. It also makes a good story, especially when park warden Myles Turner describes it all in his routine monthly National Parks report. This goes to the park headquarters, and for some reason is then quoted by Associated Press. It goes viral and gets into newspapers worldwide, and even my mother-in-law in Britain picks it up.

As Jane and I learn pretty quickly, at night there are always hyenas immediately around the house. We often hear their loud, enigmatic 'whoo-oop' call, and we cannot leave a pair of shoes on the veranda or they will get eaten. Among these hyenas there are so many different individuals that their organization appears chaotic, without, for instance, the simple territories I am used to from other mammals and birds. I first see it as a totally disorganized free-for-all. But what is curious to me at this time is that, within the apparent chaos, Solomon is not accepted by his wild kin. He ventures out, but is always in danger.

One late evening I am in bed, awake, ill and feverish with food poisoning after a rogue tin of pumpkin at a friend's house. Solomon, then just a few months old, is asleep on the veranda. Suddenly in that pitch-dark night, I hear yelps and more yelps, and recognize it as Solomon's frantic screaming. The noise moves away, out into the woodlands around the house, and fast. Barely clad, I jump out of the window with a torch, shouting at the top of my voice and following the disappearing fracas. I have no idea what wild animals may be facing me. Some 100 metres from the house I catch a large hyena in the beam of my torch. It looks at me, and drops little Solomon from its jaws. After more howls and shouts from me, the attacker lollops off, while I carry the bleeding victim back home.

There are bite wounds in his neck and back, and Solomon's jaw is broken. Applying bandages right around his head helps, and in the next few days he appears to be recovering. But after a week a complication arises. One of his large, broken back molars works loose from the suppurating jaw, next to the break in the bone. It looks horrible. I have to extract the tooth, without any anaesthetic available. Amazingly, Solomon co-operates – I hold him between my knees, and with an ordinary pair of pliers the molar comes out, with little reaction from him. From Solomon I now know that hyenas are remarkably impervious to pain – and as incidental consequence, it is also very difficult to punish Solomon physically, as he just does not seem to care. Wild hyenas, too, appear to tolerate a terrific amount of battering from other animals, such as lions or other hyenas.

The nocturnal incident also makes me aware that hyenas are cannibals, because I have little doubt about the intent of Solomon's attacker. They are infinitely hard on their own kind if an individual does not belong. Later, the hyenas' need to fit in locally, to be part of a society (part of a group or 'clan', as I call it) is demonstrated again and again when Solomon is grown up. Sadly, wherever we stay or camp in the Serengeti or on the crater floor of Ngorongoro, he is chased and bitten. Nowhere is he accepted by the local hyenas.

Jane and I are Solomon's clan. He goes out on walks with us into the bush, in between the wildebeest and zebra, and through the acacia thorns along the riverbeds. Even as a month-old youngster, he scent-marks everywhere, or at least goes through the motions while his glands are nowhere near in working

Solomon injured by a wild hyena. (Ineke Kruuk)

order – they do not produce anything yet. In such scent-marking, he drags his body over long grass-stalks so they slide between his hind legs, a behaviour that in adult hyenas enables anal glands to leave a soapy paste on the stalk at hyena nose-height.

Solomon chases anything that moves, including people passing near our house, who often are not even aware that something is chasing them until they are hit from behind in the back of their knees, sometimes floored by this large ball of hair and legs hurling itself at its victim, fortunately without biting. It is not surprising that the local people are afraid of him, having grown up with the many tales of what a hyena can do to a person, and believing hyenas to be possessed with witchcraft.

To us, Solomon is hilariously funny, but incidents mount up at such a rate that his boisterous character starts to reflect on Jane and me, in the small, sensitive Seronera community of African wardens and scientists. There is muttering. Another day, another disaster.

In the evening a message arrives from Alan, the manager of the tourist lodge, half a mile from our house. 'Could we remove our hyena, please?' It is their time for an evening drink, when tourists at the open-air Seronera lodge bar gather around the open fire. Suddenly they see this apparently wild animal walk in from the night. It may be intent only on some crisps and company, but people do not know that, and it is causing nervous customers to scream and flee, some seeking refuge on top of the bar. At other times things happen in daytime, when Solomon finds the visitors' comfortably tented camp around the tourist lodge abandoned, and he curls up in a nice soft bed in a tent. There, often he does just what hyenas do anywhere before they settle down and sleep – he relieves himself before sleeping, leaving a soaking wet bed for the lodge staff to sort out.

Kay, the wife of one of the park wardens, understandably keeps on worrying about what Solomon could do to her two small children. Other locals distrust any hyena in their midst because of the animal's association with witchcraft. But the proverbial straw that breaks the camel's back is Sandy's breakfast. Sandy Field, the Chief Park Warden, lives in a house that dominates the Seronera staff village, alone with his servants and surrounded by Serengeti savannah. Every morning his servants lay on a full English breakfast, ready for Sandy to enjoy on his veranda, with splendid views over the plains and hills. Fried eggs, bacon, sausages, butter, neat plates on a starched tablecloth, all are waiting for Sandy's entry at seven o'clock precisely. The problem is, Solomon discovers that breakfast is already there and available at a couple of minutes before seven, when nobody else is around.

Pulling the tablecloth causes everything to clatter down, and produces rich pickings all over the veranda. The subsequent angry cries and punishment are clearly a price worth paying – Solomon does not really mind being beaten with a stick. It happens once, twice, and this in the house of the most important person in the Serengeti. We are told that either Solomon is put in a cage, or he goes from the Serengeti, or we have to.

Jane and I decide that we will try to set him free, make our Solly live like a completely wild hyena, with all the others of his kind around, and far away from people. We know that we will miss him very much, but it will be the best solution for Solomon, and for our various neighbours living nearby. So we go camping again, as far away as we can go to the grasslands of the Ngorongoro Crater, 100 kilometres from the Seronera village, and we take Solomon. There are no other people around there apart from a few Masai herdsmen, and there are lots of hyenas that Solomon perhaps can live with. Often there are carcases of killed wildebeest or zebra. We hope that he will be all right looking after himself, perhaps after some initial scrapes with the other hyenas.

So one day from our camp in the Ngorongoro Crater we find a large hyena den, many kilometres from the tent, consisting of lots of holes in the ground and with many hyenas nearby, and we let him jump out of the Land Rover. We watch Solly, sniffing about with the other hyenas milling around nearby. There is lots of mutual interest, sniffing and leg-lifting, and Solomon appears to be accepted like one of the locals. Quietly we drive away in the Land Rover, leaving him to it. I think he does not even notice that we go, he is so busy. It is a sad moment but we are sure it will be best, in what seems a paradise for hyenas. Coming back to our tent, it feels very empty, we have lost our Solomon, and he is a long way away.

That is what we think. Two full days later, we are sitting in our tent along the Mungi stream on the Ngorongoro Crater floor, a lovely spot. We hear a noise and some well-known squealing, and there is Solomon back again. He is clearly delighted that he has found us, and he is desperately hungry. So we give him something to eat from a tin, and when we look him over we realize that he must have had a rough time while he was away, his coat is shabby. There are a few bite wounds near his tail, he is muddy, and altogether he looks rather

Jane walking with Solomon in the Ngorongoru Crater.

down and out. Obviously, the wild hyenas had not wanted him. We also feel miserable: Solomon's misfortune is really our fault.

Our poor Solomon. Neither Serengeti nor Ngorongoro hyenas will accept him, so we cannot just release him as we tried, we could not possibly keep him in a cage here, and we are in a terrible quandary. As a last resort, we talk to a friend of ours who has a business supplying African animals to zoos. Upon his urging and with heavy heart, Jane and I decide to let him send Solomon away. Leaving the tropics, Solomon finds a new home in Scotland, in Edinburgh Zoo. From visitors we hear that, for years after, Solomon is a very popular animal there, we get messages, and he is a special friend of the staff in the zoo.

But I still feel guilty about my betrayal of that much-loved friend. Solomon gave me a lot, he was dependent on me, I felt close to him, and with his move to Scotland I don't know what will happen to him in the rest of his life. His fate is determined by my disloyalty. Now, many years later, I feel that I should have had the guts to put him down, but in my cowardice at the time I could not do it. It is a sad ending to the life of a wild animal as a domestic pet, as happens so often. He did help me to get under the skin of those wild hyenas and, thanks to me, the zoo is his reward. I have a nasty conscience over what I did.

There is a brief postscript to this story of Solomon. Years later, Sandy Field, a splendid, kind figure and a lifelong friend, who became Solomon's nemesis and who ordered him out of the Serengeti, retired from his work as Chief Park Warden in Tanzania. He moved to nearby Nanyuki in Kenya, and continued flying his small Cessna 150 aeroplane in his retirement. One day, in atrocious weather, he crashed in the forest of the Aberdare Mountains. His remains were found only several weeks later, largely consumed by wild animals – mostly, probably, by hyenas.

Spotted hyenas. (Ineke Kruuk)

CHAPTER 6

Clans of the savannah

Ngorongoro Crater.

A FTER SOLOMON HAS GONE, we still often use our camp in the Ngorongoro Crater, along the Mungi stream. We sadly feel his absence, and look around us somewhat stunned. In the crater we are surrounded by an Africa very different from that around our house in the Serengeti – the animals are the same, but the landscape is totally other. Ngorongoro is at an even higher altitude, the crater floor 250 square kilometres of fertile grassland plain with a lake in the centre and some large marshes, surrounded by the vast high crater rim. Living here as we do is being in paradise. There are masses and masses of animals, the sounds of wildebeest grunts are never off-air, zebra, gazelle, rhinoceros and lions are always around near the tent. The calls of thousands of birds fill all the different crater habitats, with flamingos on the lake, plovers tinkering away along the streams and the sad 'ka-waaauw' of crowned cranes the call sign of the marshes. At night we have the never-ending chorus of hyena calls, from Solomon's kin, although they were his enemies.

One day soon after Solomon's departure we have a visit. Our tent in Ngorongoro is buzzed by a small Cessna plane, with a registration number I know well. It lands on the grass after I have cleared a few territorial wildebeest bulls out of the way with my Land Rover. The pilot, National Parks director John Owen, emerges from the plane with his big smile. He tells me that he just wanted to call in, to see how we were doing, and to show an American visitor around. Out of the Cessna steps his guest, the aviator Charles Lindbergh. Completely unexpected, to me this is just about the biggest culture shock one can get, meeting the man who fired the starting shot for modern travel, and all this here in the superlative natural beauty of Africa. Lindbergh, an immediately very sensitive person, is deeply impressed by being here, in our Africa.

While Charles Lindbergh and I talk about flying, about Africa, the Serengeti, hyenas and other things, John Owen stands next to his plane smoking his pipe, visibly breathing in the landscape. We take the two men around for a drive in the Ngorongoro Crater. They are two greats – and both are people who made my African existence here a possibility.

Once they are sky-bound again and leave me with all their questions, I need to think. I need to take stock in our magnificent setting, and to ask myself about my project. With Solomon now no more than a memory, I discover how I still find pleasure in how much he got me and Jane into the spirit of a hyena existence. Following on from that, meeting the two great visitors with all their questions reminded me of the importance of my main purpose to find out how hyenas live together, how they are organized and get on with each other, and to understand how they have an impact on all the animals around me. Hyenas have a highly complicated society on the African savannahs which is, as both our visitors remarked, in some aspects not entirely different from ours.

The word 'savannah' has a soft, musical overtone; it is a word that makes me happy. It is also a landscape where I feel at home. I like being in forests, where I can watch badgers, or in dunes or mountains, or near water when I am watching otters. But the savannah has something more basic that attracts. It has horizons, it has trees on endless grasslands, with a massive diversity of

Pack of spotted hyenas.

wildlife. Yet, although the number of species of mammals, birds and plants may be second to none, that still is not for me its fundamental attraction. Of course it could be my upbringing in Holland, a flat country if ever there was one. Rather, I believe that it is a deep-seated, primeval instinct accounting for my contentment here, in a habitat where people evolved. The savannah cradle of mankind, reaching deep into the millennia of evolution, is not far from the Serengeti, both in a geographical and in a physical sense. Olduvai Gorge, where the Leakeys found so many fossil human remains, cuts deep into the Serengeti plains.

Many animals appear to share the same predilection for the grassy, open plains with their flat-topped acacias as their natural home. Also, a pride of lions, a single elephant or a herd of gazelle, to us they all look better on the savannah, more magnificent than anywhere else. On the face of it, to be a researcher here who is concentrating on just one species is being small-minded, almost sinful. Yet that is no concern to me, because when I am out on the savannah, I feel completely open to all this country offers apart from my own project. Somehow, concentrating on one species makes me try to see everything around here through the eyes of that one animal – I have thoughts such as, 'If I were that hyena, where would I go?', or 'What would I do?' Having had our tame hyena, Solomon, is a great help. But what I am really after, of course, is the wild animals that surround us.

At any time of day, somewhere near our Seronera home that we call the 'hen-house', I may from my vehicle watch a single, well-rounded spotted hyena crossing the beautiful Serengeti expanse, walking in its somewhat doleful gait, with Kimunya, one of the small hills, providing part of the background. One wild hyena, all on its own.

But is it alone? In my early days here, I can make little head or tail of these animals' social connections. With my previous experience I am used to seeing many other animals as singles, or in pairs, in packs, in prides or in colonies. For the social picture of spotted hyenas I can only see total chaos. Whenever I am with one lone individual, before long it will be together with numerous colleagues, seemingly well acquainted with each other.

Whatever their social system is, it is evidently a very successful one, given the large number of hyenas in these parts. It is a system that rejects an incomer, such as poor Solomon. The spotted hyena in the Serengeti and Ngorongoro is a numerous species. I want to find what their community life is, if they have any at all. But to make sense of their social organization I need to be able to watch wild individuals properly, over long periods, and so far I have little luck with that. Later I will find that a tremendous strength of the hyena social system is exactly that what causes my initial bewilderment, the fact that sometimes they are alone, and sometimes they co-operate with others in packs.

In my first few months in the Serengeti I still find it difficult to drive around cross-country at night. I just do not have the confidence, often not managing to avoid the many huge warthog holes and damaging the car. I get stuck in the *karongas*, the riverbed crossings, I am worried in the dark about meeting elephants, rhinos and buffaloes. I feel incompetent about it, and in my mind it is becoming a battle between me and the hyenas.

Minor aggression within a hyena clan.

Eventually, as time goes on I find some confidence in following animals by vehicle on their nocturnal wanderings. I realize that what is slowing down my studies here is not only my initial lack of African fieldcraft, but also simply inexperience in the context of the vastness of the Serengeti, of the enormity of the plains, with the vast animal migrations that submerge a single area under a black mass of wildebeest and zebra one day, leaving it totally empty the next. In these early days, the Serengeti is just too big for me.

I am not the only one who meets this problem, as some of my zoological colleagues here also feel themselves up against it in the Serengeti. The couple who are studying zebras, Hans and Ute Klingel who are living nearby, advise me to follow their example and start watching some of my carnivorous subjects in a much easier area not too far away. It would enable me to acquire some first and basic experience, before tackling the Serengeti. It is their advice that first took us to the Ngorongoro Crater, a caldera which is like a compact, mini Serengeti. But I am determined to also keep up the efforts on the wide Serengeti plains, and throughout the following years we often move between open plains and the deep caldera.

The Ngorongoro Crater floor is effectively an entirely isolated world, although with some difficulty animals can come and go over the rim. From the grassy plain at the bottom of the crater one sees the rim all around, often topped by white clouds spilling into the caldera, like milk spilling out of a bucket. There is a large lake in the centre with flamingos, there are tall acacia trees along the streams, and there are small hills. And there are masses, simply masses of wild animals; it is a hugely fertile area.

With its large herds of grazing wildebeest, other antelopes and zebra, the animal species are much the same as in the Serengeti. There are small differences, such as that Ngorongoro has no giraffe, animals that cannot cope with its steep, high rim. But for me, the most important zoological feature is the great number of spotted hyenas, which are not quite as nocturnal as they are on the Serengeti plains. In Ngorongoro all species are less shy of vehicles and people than they are in the Serengeti. What more can I ask for?

On the crater floor Jane and I camp, or when possible we stay in one of a couple of small cabins. We go there frequently, for weeks at a time. When camping, we keep all our provisions in the open front of our large tent. One night I wake up in the depth of darkness, because someone is moving about amid our supplies. I shine a torch and stare into the bright eyes of a wild hyena, a couple of metres from my sleeping bag, right inside the tent. I shout at it, at the top of my voice. It is a pity to have to chase it away, but the animal is off – I almost feel that I have to apologize to it. However, my frightening actions are less effective than I think. Next night, the hyena is back again, almost certainly the same animal.

When it turns up a third time, I take it more seriously with a decision to mark it, so I can recognize it in daytime. With a dart gun from my sleeping bag, I immobilize it, and remove a small triangle from the edge of its ear. Meeting it next day, amid a crowd of its mates, the hyena shows no rancour. The event also shows that clips from the ear appear to be a useful method for individual recognition, which is a prerequisite to sorting out their social ties.

After that I mark many more hyenas, though later I really regret it – the darting and marking must have caused suffering, and it would have been better to try to recognize individual hyenas by their spot patterns alone. Alas, there is no way any more to apologize to the animals. And although I am sorry to have used it, the method does open up ways to expose the hyena's social life, first of all here in the Ngorongoro Crater.

Once a hyena is darted (usually in the rump) and the drug starts to take effect, the animal begins to show the first small signs of disorientation. After a few minutes it begins to stagger, and then it lies down. This is where I have to be very careful – as soon as there are signs of abnormal behaviour, other hyenas viciously attack it and, suspecting what they can do to one another, I have to head them off.

Afterwards, I can see this observation as a kind of (unplanned) experiment. It shows that being ill or injured and behaving differently is a fatal failing in hyena society. They are not like elephants or whales, which may help each other. It gives me a clue about why Solomon was rejected by the wild hyenas. We humans see it as callous behaviour, but it is not confined to fierce predators such as hyenas. Even the gentle antelopes show similar disdain for anyone who steps out of line. Wildebeest, one of the most gregarious animals on the African plains, have no time for conspicuous individualists in their midst, as I find out sometime later.

Talking to my Serengeti friend Hugh Lamprey about my work with hyenas, he mentions that, many years ago, a former colleague from the Game Department had been interested in where the migrations of the famous, enormous herds of wildebeest went. This was a very difficult question in days before aeroplanes were used extensively in the national park. To find out, he darted a couple of

Hyena female carrying off her share from the kill.

wildebeest and put a broad white collar around their necks, hoping to recognize them at a later stage. However, within days both collared animals were killed by predators, one of them by hyenas. Hugh thinks that these conspicuous wildebeest were selected out because they looked 'abnormal'.

Years later I decide to test this idea. Staying in our small hut in the Ngorongoro Crater, surrounded by its very high density of hyenas as well as wildebeest, I am aiming to try to make some wildebeest stand out from the crowd to see if they are more vulnerable to predation. With my colleague Patrick Duncan, we dart and immobilize 30 yearling wildebeest, and provide all of them with a small, coloured ear-tag. In addition, half of these very black animals get their black horns spray-painted to a horrible, luminescent yellow. Fortunately for them, they cannot see their own horns. The other 15 of the darted wildebeest remain unchanged in their appearance. Over the next few days I watch the Ngorongoro wildebeest, expecting predators to single out the yellow-horned ones.

What happens is much more surprising. On two different occasions I see one of the yellow-horned animals, both times vigorously pursued by several other wildebeest, and not allowed anywhere near the herd. On one occasion I see one of the ear-marked 'controls' with ordinary black horns, happily grazing with the others. The observations are few, and I realize that I am hampered by my small sample in those large herds of tens of thousands of black animals.

Alas, I do not detect any predation on the marked animals; my sample is too small. But what I do see strongly suggests that wildebeest eject from the herd colleagues which lack the uniformity of the crowd. I believe that it is exactly such intolerance that will expose a potential victim to predation. I could speculate that it also would prevent a yellow-horn gene, if there were any, or any other gene that flags an individual as different, from spreading in the wildebeest population. Whether you are a hyena or a wildebeest, or perhaps any other social mammal, it does not help you if you are different from the others of your species.

Although the observations are not enough to get a definite conclusion, they at least suggest that 'sameness' is an important protection for the wildebeest. It is colour discrimination if ever there was any. Probably it is the same for many other animals too, including hyenas. If an animal is different, it could endanger the population, especially if it would breed.

Another astonishing kind of aggression that I see in hyenas amounts to warfare. By first light of day, Jane and I are driving slowly through the morning mists of the Ngorongoro Crater plains. It is cold and the sun is not yet above the rim, but there are flocks of egrets, a stork, the air full of the 'kawauw' calls of crowned cranes. Wildebeest and zebra are grazing close to the marshes, a blacksmith plover tinkers along a puddle. I drive to the top of a small hillock where we sit on the bonnet of the Land Rover with our binoculars, soaking it all in, without having to talk. Next to the car is a large pyjama-lily in flower, white with pink stripes, and further out on the grasslands are several bright red *Haemanthus* lilies.

We can see animals everywhere, and for kilometres. Scanning the land around us is an excellent way to get the feel of the country, and for us to begin to see what is happening in the hyena world. The few we can pick up on the grasslands appear to be asleep. Below us and not far away is a large hyena den, a cluster of holes in the ground with patches of sand around them, and right now with just one large, fat female asleep between the entrances. I know that those holes are used by many. I find several such large dens in the Ngorongoro caldera, seven in all. One of them, at a place called Oldonyo Rumbe, has tunnels as big as manholes connecting to some underground drainage system, tunnels so big that a person can easily crawl in and out. I have to try it, of course, and come out again covered with hyena fleas.

Over time, I find that all spotted hyenas belong to groupings, or 'clans', each centred on one of these major dens. The clans have a highly complicated system of communication, or 'language': they are very vocal, noisy with more than a dozen different calls and masses more intermediates, including 'whoops', 'yells', 'howls', 'grunts', 'staccato grunts' and many others. Each call has its own meaning of aggression or fear or warning or attraction. Clearly the clan members know each other, and Solomon would be an unknown intruder, to be expelled. For any carnivore such large groupings, of up to 80 individuals, are

Battle between clans.

an unusual arrangement. Most carnivore species live alone, or in small groups, packs or whatever they are called.

Hyena clans indulge in dramatic battles, involving large numbers of individuals, in fact almost complete armies. The organization of these animals is unusual at every turn, and perhaps especially interesting when so often one is reminded of people's own affairs.

Konrad Lorenz, a world-renowned Austrian animal behaviourist of the twentieth century, once wrote: 'They may defend a territory or food. They chase, and fight. However, animals do not kill others of the same species, and of all living beings, it is only mankind which indulges in large-scale warfare.' Perhaps it was his love of nature that made him perceive natural life to be as rosy as that, but wrong he was. Animals often indulge in behaviour that, in our own species, we would call evil. There is murder, rape, theft, warfare, and cruelty against the infirm or against any individual that is different. It is all there.

On our small hilltop, Jane and I drowsily wait for things to happen among our hyenas. The air is quiet, wildebeest and zebra graze in the dew. Then something stirs. At some distance from us, a few small groups of wildebeest start to mill about, instead of quietly grazing as they were doing before. I notice that a single hyena is running through them, fast, obviously not interested in any of the animals around it. It runs in a straight line, at full speed and with its eyes on something far ahead. It comes closely past our Land Rover, heading for the long-grass, hilly country near the rim of the crater. Clearly, it senses some important business there. Only then do my ears, too, tune in to a distant excitement, faint and barely audible but clear: it is the whooping and yelling of many hyenas. It explains the absence of hyenas here in the low hills around us: the animals are away en masse. Immediately Jane and I are off in the vehicle, following the single runner across the grass, bouncing over a long distance.

Cresting a rise, I look down on the other side where we are confronted with many hyenas, and chaos reigns. There are remains of a recent wildebeest kill, probably part of the history of what is happening, but right now this carcase

Part of a hyena clan on kill.

Pack of hyenas feeding. (Ineke Kruuk)

is not the centre of attention for the 40-odd hyenas. Small parties, with tails up, chase others with tails between their legs, with their peculiar, hysterical 'giggling', loud whooping and yelling – a very human-sounding madhouse. I see a chase over a couple of hundred metres before the parties turn, with chased becoming chasers. There are scraps between individuals, chases again, with groups of hyenas standing still and filling the air with deep, melodious whoops, *basso profundo*. One animal takes a few bites from the wildebeest carcase, another carries off a leg and is chased until it drops the bone, yelling.

I realize that the furore is between two of the hyena clans, the Scratching Rocks clan and the Mungi clan, named by us after their respective dens. We are in the region somewhere between their two territories. From their markings I recognize quite a few of the participants, males and females, fighters and bystanders. Heaven knows who started it, and who killed the wildebeest. But whatever happened before, the two clans are in major battle, and things have clearly gone beyond a little scrap.

Suddenly a party of six of the warriors manages to pin down one hyena from the opposition. A heap of fighting bodies, with growls, yells and howls; more hyenas pile in, while others watch excitedly in a somewhat distant circle. From the latter, the spectators, the ones I recognize are all Scratching Rocks members. Of the heap, presumably all from Mungi, it is impossible to see details, and all the participants are pretty dirty by now, many bloodstained.

Five minutes later the Scratching Rocks spectators attack en masse. This time the Mungi hyenas leave, run and keep running, yelling, and for now seemingly for good. They leave a single victim lifeless, a young male, now a bloodied corpse with some parts of the body bitten off. Later I notice that nobody eats the hyena carcase until the next day, when two hyenas reluctantly take some soft parts, now decomposing. It was not cannibalism that drove the orgy I watched. It was territorial warfare.

I realize that in such a fracas our tame, solitary Solomon, would not have stood a chance. In a way it is an insight I am pleased about, because such wonderful observations of wild animals do not come every day. In time,

Hyena pack on boundary latrine. (Ineke Kruuk)

sightings such as this one build up to a fantastic picture of hyena society, of an organization that would not look out of place even for mankind.

Over the years, I come to recognize many of the 300-odd hyenas in the Ngorongoro. I know their clans, and I know where they can go on the crater plains, and which areas are taboo for them. Their life is not a free-for-all, and there are strict boundaries. Each of the seven hyena clan territories in the Ngorongoro Crater has well-delineated borders, not obvious to us but marked by latrines, and known in detail to the many animals involved. Each clan counts scores of members, some as many as 80. The present two, which I am watching here in battle, each number about 40 hyenas. Within the clan, hyenas frequently check on each others' credentials, in conspicuous 'meeting ceremonies'.

Later, I find that territorial border incursions often happen 'by mistake', when they are hunting and their wildebeest quarry crosses a boundary as it runs next door to escape. Hunters may pull it down just across the boundary, where ownership is then contested by the rightful territory holders. Some borders are abundantly scent-marked with latrines. Often I come across 'patrols', parties of hyenas led by a female, and which follow, check and scent-mark the boundary lines. As a hyena, and especially if you are one of the neighbours, it is better not to get in the way of such a patrol.

We return from Ngorongoro to our much-loved 'hen-house' in Seronera with a load of new experiences, with data, and with a new appreciation of hyenas.

For a researcher like myself, hyenas are just one of the attractions here. Staying overnight in Ngorongoro Crater is an experience dreams are made of, because of the beauty of the landscape, those magnificent early mornings in the cool mountain atmosphere without other people, the thousands of animals, roaring lions, the huge fig trees with hyraxes, the flamingos on the lake. Other scientists here study rhino, wildebeest or zebra. For our little research unit in the Serengeti, Ngorongoro becomes a kind of laboratory for the Serengeti.

After my return to Seronera, I talk to an important meeting of trustees of the Serengeti Research Project, consisting mostly of grandees from African government departments, universities and other institutions. I lecture them about what hyenas eat and how they set about it, how important they are in the ecosystem, and about their social life. In the trustees' eyes, there is interest, but also some disbelief. Ideas about these animals are deeply entrenched. One small problem arises when I tell them about my diet analyses from scats, from faeces – and trouble arises in that the Swahili name for hyenas is *fisi* (pronounced 'feesee'). I can see the question marks on the faces of my audience.

During an interval in the meeting, Louis Leakey, one of the trustees, the grand old man of archaeology who discovered the African origins of mankind, shows the assembled scientists and trustees a tame bull wildebeest called Oliver. The animal is free-ranging around the laboratory site near Seronera. It is an individual that Leakey had donated to the research scientists a few months earlier, for grazing observations. Oliver had attached himself to him as a calf, when Leakey was working in the Olduvai Gorge through the Serengeti plains. He wants to demonstrate the animal's obedience to him, with his repeated commands of 'Oliver, lie down.' After a few of such, Oliver takes one good look again at Leakey and charges, flooring the illustrious scientist amid great hilarity.

During that meeting with the trustees, when I tell them about my questions and some of my results, I do not talk about the intense pleasures of what I am doing, about the wild excitements I find in the hyenas beyond anything I had done before. But I do tell them about Solomon, and I watch the Tanzanians shaking their heads in disbelief. A hyena, as a pet? Prejudice about hyenas is deeply engrained here. These important people, with their wide experience in fields far from mine, are unimpressed by what I see as one of my major findings, the hyenas' clan system, the organization that allows the animals to behave as co-operative groups as well as loners. It makes them successful in coping with all kinds of different prey, with anything from a flamingo to a zebra or a buffalo. It makes them the most abundant large carnivores in the Serengeti.

CHAPTER 7

Hyenas hunt

Hyenas hunting zebra at night.

EVEN WHEN OUR tame Solomon is only a few months old, I notice his passion for legs, especially when they walk away from him. He doesn't stalk, he goes for them without ceremony, flat out, whether the legs are human or animal. It is the hunt of a youngster. But it is a hunt all right, a behaviour that in wild hyenas is central to their role in the Serengeti, in the ecology of animal society. And whether watching Solomon or the wild hyenas, I realize how much of a hunter I am myself, from the excitement that it also arouses in me. All the more so because with the hyena's hunting I am discovering something totally new. Hyenas are not known to be hunters: this is new ground.

Moreover, hunting is the one habit that, perhaps more than any other, brings hyenas in conflict with people, as well as with lions. The animals are not just targeting any old or small prey, but they are hunting exactly the same species that people and lions are also most interested in. And in nature, whenever two species share a space and are interested in the same kinds of food, there is trouble.

Going as fast as I dare across the Serengeti plains, I am bouncing wildly inside the Land Rover pickup, desperately trying to avoid holes and termite hills. The day is yet young, Africa is just stirring and full of promise, but I have eyes only for the grassland ahead and the two figures rushing alongside me and in front, a large wildebeest closely pursued by one single hyena. I can feel my adrenaline flowing, I want to be there at the end with the hunter, never mind the bumps and the risks to the vehicle.

Others in the grazing community stay well back: other wildebeest, zebra, and gazelle, all step aside and stand there rubber-necking at the drama that is taking place. There is an empty corridor ahead of the wildebeest bull running through the large herd, no one steps in to take his side, it is only him for himself. I am almost overcome by what is happening here, seeing a large life at the point of being terminated.

However, after a run over about a kilometre, the wildebeest's prospects slowly improve. The gap between it and the hyena increases. The hunter stops and looks around him, as if to state that he did not mean to get that wildebeest anyway. It is just one single hyena, which had been chasing after one, much larger, wildebeest bull. I am breathless, too, and stop the engine to contemplate the world around me.

The episode reminds me of a scene in Oxfordshire in England, with riders on horseback in full gallop, with pink coats and the sounds of a hunting horn and tally-ho, a pack of hounds and horses during the chase of a fox. Here in Africa I know that such events, the chase and the kill, are universal in the natural world. I know that some creatures, the hunters, chase and exhaust and kill others. But also here, as in England, many a quarry escapes, probably to be hunted again at some later time. Life in nature is tough, and animals do not die in bed.

Although they remind me of fox hunters, these spotted hyenas are more like wolves in the way in which they search and take their prey. They do not get involved in some 'cowardly' scavenging act, but this is hunting, the real thing.

Hyenas killing wildebeest, after a chase. (Ineke Kruuk)

Another day, another drama, this time late in the day in the Ngorongoro Crater. A single hyena walks across the central grasslands, through herds of hundreds of wildebeest and gazelle, which seem to take hardly any notice. They barely avoid the hyena, allowing it to come within some ten metres. Here in the crater there are so many hyenas walking about, and a wildebeest cannot stop grazing for every small interruption, it happens all the time.

But this particular predator means business, clearly taking an interest. Every so often it stops and looks at the animals around it, at the wildebeest that are grazing or just staring. It walks on again, seemingly without any aim. Suddenly the hyena begins to run towards a dense part of the herd, some 30 metres away, creating quite a disturbance. All the wildebeest rush off and the hyena stops, head up, surveying the panicking black hordes. This time there is nothing to attract it, it loses interest and walks on. But after the next such test-rush, the attention of the assessing hyena is drawn to one particular running animal, a wildebeest bull.

There must be something in that particular wildebeest's gait or other behaviour which I cannot see, but the hyena can. It marks out that one individual as vulnerable. Some people can assess racehorses by just seeing them move – I do not have that ability, but hyenas can do it with wildebeest. The hyena I am watching takes off and runs after that one bull, flat out. The chase clears a path through the herds, and the bull loses his previous anonymity within the black crowd – it now stands out. After almost a kilometre of running, a grip of the hyena on its quarry's hind leg stops the wildebeest. It stands, sways its head, and moans.

Soon a dozen other hyenas, having previously been asleep until the commotion, join the fray from different directions. The victim falls, the 12 of them eat their full, and an hour later the carcase is dispersed. It is another one gone, leaving no more than a black patch on the grass.

Hyena carrying off the spoils of their wildebeest kill. (Ineke Kruuk)

Hyena pack with wildebeest yearling, after a chase.

Not far from where this drama takes place, two Masai herdsmen are walking a herd of cattle across the Ngorongoro Crater floor. It makes for a classical African picture, the tall herders with their spears, wrapped in dark ochre-red, with their 50-odd cattle ahead of them. The Masai take no notice of the hyenas and their doings, though the hyenas clearly keep an eye on them for they are alerted by anyone walking.

The crater, these days, is not a strict nature reserve (it will become one later), but now it is shared between the Masai graziers with their hundreds of cattle, and the wildlife in its thousands. There are two Masai villages here, *manyattas*, hut circles with a thorn fence where at night people and livestock share. The wild animals do not bother the people much, and there is little conflict. That comes later.

The Masai people go their own way, we go ours, and I do not see the Masai particularly interested in what we are looking at. To them, wildlife is a decoration of their day-to-day environment. For me, it is an aspect of that environment that I dearly want to know about.

Another day, another experience. I am watching the hyenas again, sitting in my Land Rover, keeping an eye on an old friend. With age, she has rather lost her figure. In fact I, as an often unkind observer, might call her obese, with her skin drab, her lips flaccid, chunks missing from her ears. She is not a beauty, to put it mildly, yet it is she who calls the shots in the Scratching Rocks clan here in Ngorongoro Crater. Without any ostentation, she is the centre of this spotted hyena society, one of their female warriors, the Amazons. Classical antiquity, the Greeks, knew the Amazons as the feminists par excellence, ferocious female warriors. This particular animal is scientifically known to me with a number, but usually Jane and I refer to her as Ute.

It is early evening, and a group of nine hyenas are still asleep, scattered on the grasslands but clearly belonging together, at quite a distance from their huge den. I know that the clan has some 40 members, most of them at the moment indulging in solitary slumbers throughout the territory. The light is fading fast, and every so often a hyena lifts its head, then subsides again. There are soft grunts of wildebeest some distance away, some flamingos fly over, a vulture circles.

Ute gets up and just stands with her head drooping, hardly moving at all. Another hyena moves, then she and Ute sniff each other, and the second one lies down again. Others are also beginning to move and there is a general slow awakening, aimlessly ambling and being aware of each other. Night falls.

For me, this is one of those exciting times in my life when I am witnessing something in nature that completely overturns our understanding and, in this case, our knowledge of what is in any case a large, fascinating wild African mammal. It is part of my observations here of events that totally alter our perception of hyenas.

Ten minutes after the party started fidgeting, Ute slowly walks away, taking little notice of the others. The remaining eight also get up, and follow. They walk as a band, with apparently little leadership, yet somehow the pack is directed. It is the females who appear to lead; in spotted hyenas they are larger than the males. They are the Amazons.

It is almost dark now, but with the strong moon I have no problem in keeping up with the animals in my Land Rover. I rarely even have to use my lights. None of the hyenas seem to take any notice of me. They are heading into a hilly part of the enormous Ngorongoro Crater floor, away from the large wildebeest herds that graze on the flats. I glance at the silvery, undulating landscape, again experiencing Africa at its most beautiful.

After half an hour's drive next to the hyenas through fairly long grass, I see a zebra family looming up ahead. Through binoculars I distinguish a stallion, five mares and three foals, all of them now getting restless. The animals' heads are up, the hyenas walk towards them in a slow, steady and almost purposeful

Hyena pack at night, looking for zebra.

way. They stop some ten metres short; both parties stand and look at each other. Then slowly the zebra stallion, holding his head low, walks out from his family, towards the hyenas. He puts on some speed, and in no time the hyenas turn and scatter out of his way.

Rejoining each other, the hunters walk on, and before long I see them approaching another zebra family. This time nine of the hyenas speed up. The zebra are seven in total, a stallion guarding his mares and foals, their loud, high-pitched barking calls rending the night air when they become aware of the danger. A short chase follows, with the entire pack of hyenas bearing down on the family. The stallion is a foe to be reckoned with, running behind his charges, biting and kicking at the hyenas.

Briefly, it seems that the stallion has seen off the threat to his family. He appears to be everywhere at once, hyenas scattering in front of him, then returning again, jumping out of range of the zebra's flailing legs and snapping teeth. But within minutes two hyenas take a grip on a leg and a loin of the stallion, slowing him. At the same time, several other hyenas attack an almost fully grown foal, surrounding it and biting it anywhere. There is confusion and ear-splitting barks from all the zebra. The stallion breaks loose and follows the remainder of the family, and the foal is left to the predators.

Early next morning I find Ute, her head bloodied, fast asleep near the den. She is bloated with food, and I would not be surprised if she had eaten most of the maximum quantity a hyena can stow away in one feeding session, which I estimated at 14 kilograms. Later that evening, and in that very same area, a pack of 12 hyenas again follows menacingly behind a family of zebra. A similar battle develops into a headlong flight, leaving one of the large adult mares standing alone in a seething crowd of hyenas. Another victim.

After Jane and I move back again to the Serengeti, we are out during a wet season daybreak and thousands and thousands of Thomson's gazelle are scattered over the wide grasslands, close to our house. In my Land Rover I am still musing over last night's visit to our house of a genet cat, which just walked in while we were talking. It was attracted to flying termites that were accumulating under the lights, a beautifully spotted animal with a long, ringed tail, taking little notice of us. This place is so incredibly rich in animals, in carnivores and many others, in birds and insects. I drive past a cheetah on a termite hill, a kori bustard booms with his huge throat balloon.

A shiver runs through the gazelle-studded plain in front. Animals run, black-and-white flanks line up together, many are 'stotting', fast high-jumping on stiff all fours. Through my binoculars I watch a lone hyena racing after a male Thomson's gazelle, the hyena at full speed. The tommy does not appear to run flat out, but stays just in front of the hyena. I cannot help being surprised about its apparent lack of effort, because if I were on its hooves, I would go as fast as I could possibly make it. As the two disappear over a slight rise, the odds are against the gazelle.

Over the years, I see many hunts like these in the Serengeti, and in the Ngorongoro Crater. Here I describe it in factual terms, but even after having seen it so often I cannot help but feel strong emotions when it comes to the end of these large animals. Years later I still have that sense of slight horror

Hyena in herd of Thomson's gazelle, looking for fawns. (Ineke Kruuk)

when thinking of some of those events I saw in Africa. Almost forcibly I have to remind myself that what I see here on the plains, this to me ghastly way of dying, is the norm for wild animals. It is still difficult not to be anthropomorphic about it.

As a scientist, I also see the fascinating, hard statistical facts behind these deaths. First, spotted hyenas come through as predators such as wolves, and their image of a scavenging nuisance flies out of the window. In fact, in Ngorongoro Crater hyenas are the most important predators of all, never mind the lions, leopards, cheetah and wild dogs. And, I would almost say, there is a sophisticated organization underlying their hunting.

When hyenas hunt zebra, they operate in a pack, quite a large group of often ten or more, with a female lead. It assembles well before the hyenas make any contact with their prey: they have 'decided', long before they get there, that today is zebra-hunting day, not wildebeest or gazelle. What makes them decide, I do not know. Experience from the previous days' hunting, perhaps?

Judging just by their social hunting, it is the big females, the Amazons of this world who, without ostentation, keep food coming in, who lead the big hunts that target the zebras. In contrast, if hyenas chase wildebeest or gazelle, any one hyena, male or female, may just start on its own. Sometimes two or three of them join up, but only after the end of the chase do larger numbers of hyenas assemble to take part in the feeding frenzy.

Hunting is a striking piece of organization by the hyenas, with zebras being tackled in pre-assembled packs, wildebeest or gazelle in solitary pursuit. It is efficient, because zebras live in families that are vigorously defended by a stallion, and it needs an entire hyena pack to penetrate through this defence. In contrast, although wildebeest graze in large herds, they are on their own as soon as they are confronted by a predator. For wildebeest, it is each for itself: the hyena catches the individual that stands out, or the last one from a fleeing herd.

Driving across the Serengeti plains with Jane and with Niko Tinbergen, the smell of endless herds of wildebeest hangs in our nostrils. The grass is green, the January sky of the wet season shows towering clouds over herds of zebra, gazelle and hartebeest. Wildebeest scatter in front of the vehicle. 'They're like little outboard motors,' Niko comments, as the light-brown calves run very close alongside their black mothers. Even when only one day old, calves reach impressive speeds. There are masses of them in the large herds, all born within a few weeks of each other. I stop.

All three of us are looking at two hyenas that slowly amble across the grass, their heads up. They show that they mean business. Wildebeest mothers with calves are moving to the other side of the herd, but otherwise not one of them appears to pay much attention to the two predators. When the hyenas start running we realize that they have seen something, one particular mother with a calf, perhaps born more recently than the others, but to us not obviously so. As always, the hyenas have an eye for it.

Soon the two hyenas catch up. The wildebeest cow slows down, lunges at the attackers, who easily avoid her. It is only after a whirling fight of a couple of minutes that one of the hyenas brings down the calf, the mother distracted by the other one's bites at her own legs. If a lone hyena had attacked, its chances would have been much smaller, because the wildebeest cow's horns are quite effective and the predators have respect for that. But with two of them, the mother had no chance. After seeing many such attacks, I now know that two hyenas are five times more successful than a single one, when hunting calves.

Co-operation is one of the great advantages enjoyed by social predators such as spotted hyenas. Together, they can do things that would be quite impossible for a singleton. They show it again in an immensely impressive performance on a visit I make to my Ngorongoro hyenas, when I come with camera people to make a hyena film for National Geographic.

One evening with the Scratching Rocks clan of hyenas in the Ngorongoro Crater, I have my Land Rover parked close to their den and the film crew is in their vehicle behind me. A dozen-odd hyena clan members are scattered on the grass nearby, asleep. Everything is dead quiet.

Unexpectedly, at almost a kilometre distance we see a large black rhinoceros come into view, with a small rhino calf at her heels, and a cattle egret sitting on her shoulders. Slowly, almost nonchalantly, the party comes in our direction.

Just caught a wildebeest calf.

The mother is a huge, tank-like animal, with small, menacing eyes, large ears, and a very impressive horn. Rhinoceros eyesight is poor, so she can see little of what is waiting for her here.

All of a sudden, without any warning or provocation, there are about 25 hyenas around the rhino. Tails up, darting backwards and forwards towards the calf, with the enormous, ponderous mass of the mother whirling at dazzling speeds, attempting to horn the hyenas, managing to throw one of the attackers twice. The calf squeals as a hyena grabs an ear and lets go again when the mother interferes, another one bites at the calf's tail, others again bite at the ears. The calf attempts to hide underneath the mother, but finds little solace under the fast twisting movements of the huge maternal bulk. Hyenas are everywhere.

Unbelievably, this battle goes on for a solid 2½ hours, well into darkness. I feel breathless from just watching it, and there is no doubt in my mind that the rhino calf has little hope of survival. The attack by the hyena mob will get the better of those two, despite the efforts of the large bulk of maternal aggression. It is the power of a group in action. Then again something quite unforeseen happens.

Out of the night a wildebeest bull comes running in, right past the den where we are, with one single hyena hot on its heels. Immediately and without any hesitation, the entire hyena mob goes after the wildebeest and joins the single hunter, obviously changing their bet. The rhinoceros and her calf trot off at speed in the opposite direction. The calf is seriously injured, but it is able to run all right. The next morning in daylight I see it again with its mother, quite sprightly though damaged. I do not want to speculate how much longer it will be able to live in that area.

Confronting a black rhinoceros with calf.

The entire episode, seeing hyenas faced with such a huge, well-equipped and aggressive opponent, showed the potential and enormous advantage of group hunting, even though the rhino hunt was unsuccessful. With their clan life, hyenas can count on having others to help, yet at the same time still maintain the possibilities of going hunting on their own where that would be more profitable. It is an organization that in animal societies is quite rare, but to us humans it sounds pretty familiar.

What all this shows, also, is the effective use the rhinoceros can make of its sole weapon, the horn. Conservation authorities in Africa remove the horns from rhinos in areas where poachers threaten their existence. Rhino horn sells for sky-high prices in Asia, and in Africa current poaching levels are such that the species cannot survive much longer. Wildlife managers, in the belief that poachers will not be interested in a rhinoceros without a horn, immobilize rhinos and saw off the animal's weapon – but little do they realize that this robs the animal of its only effective defence against predators. Clearly, it would be better to have protection from more wildlife rangers in the field than to disarm the rhino.

Soon after this episode I realize that the rhinoceros of Ngorongoro Crater is in even more serious trouble. There are only a few of these huge, harmless animals left here, and one morning I am all the more horrified to find, only a kilometre from where I watched the hyena battle, a large bull rhinoceros dead, with a long Masai spear sticking out of its flank. It is immensely sad to see the enormous body lifeless. What a tragedy, what a waste.

The Masai, until then sharing the grazing and fresh water in this fertile area with its wildlife, have done so since time immemorial. They have now been told by the management of the conservation area that they have to leave

Mother black rhinoceros defending her calf against a spotted hyena.

the Crater. They have to find grazing somewhere else, leaving the crater floor to wild animals and tourists. In protest, Masai warriors take out their wrath on one of the main attractions for visitors, the rhino. Many animals get killed, and in only a short time very few black rhinos remain in that incomparable landscape. After this, every rhinoceros in the crater gets its own armed guard.

It is a sad indictment of wildlife management: Africa is in desperate need of toleration. Opposite interests can live together – witness the lions and hyenas here, where either one takes from the other and there is great animosity between them. They live in perpetual conflict. Yet both do survive, and in abundance. People and wildlife should be able to achieve something similar.

I return to the Serengeti from the Ngorongoro Crater with a head full of memories, and with a notebook full of data. Going about my business alongside the Ngorongoro Masai, often within view of their *manyatta*, I have no problems with them, Jane and I sometimes help them with a lift, or a drink of water, or with some medicine. And their country gives us insights, knowledge, beauty, and a wonderful time.

Thinking about what I learn there, it is an exciting discovery that the clans of spotted hyenas are dominated by the Amazons, the large females that evoke the nation of female warriors in Classical antiquity. Amazons lead the hunts, and Amazons lead packs going into battle with neighbouring clans. I am also struck by the highly complicated system of communication of these spotted animals, with an elaborate vocabulary. There may be as many as 80 individuals in a clan, but they all know each other. They know each other's individual voice, as I can see from their reactions when a clan member calls from a distance.

To me, the long drawn-out 'who-ooooooop' of the hyenas across the Ngorongoro night is the most evocative sound I will ever know, uttered by the callers with their mouth close to the ground, while quietly walking along. This, and their vocabulary of many other calls, the giggles and grunts and screams and growls and all kinds of intermediate ones, are an essential component of the animals' clan system, which itself is a beautiful adaptation to the hyenas' exploitation of those large herds of African ungulates. Sometimes lone hunters, sometimes in co-operating packs, theirs is a system that demands excellent communication, rare if not unique among carnivores.

When I started my work in Africa, little did I realize the excitement of new discovery that was to follow, of the nature and role of one of the most dramatic hunters of the African wilds. And all this because of the simple question to me from the director of the National Parks, who was worried about a large-scale game-culling scheme that was planned by the government of Tanganyika.

CHAPTER 8

Witches, and death in the dark

The threat of evil.

ANYONE CONCERNED WITH conservation is concerned with the relation between mankind and animals. There is no continent with such a rich inheritance of this as Africa, and in Africa, as I am finding out, there is no animal providing such rich pickings of human relationships as the hyena. When watching these animals, one cannot avoid noticing what happens between people and hyenas. At its most simple, there is the role of hyenas in the national park, in the ecology of the vast masses of animals that use the grassland plains. As their most abundant predators, hyenas can prevent overgrazing. In contrast, and just as important as far as we people are concerned, there is the negative side to hyenas, their nuisance to pastoralists and farmers.

All this involves rational arguments about an enigmatic animal. But there is one totally different aspect to the relation between hyenas and people, something very irrational and weird, hugely supernaturally important in people's lives out here in the bush. Witchcraft. Hyenas are deeply involved in this, much more so than any other animal, and over the years I come across strange beliefs many a time.

I love living here in Africa. I romp about with my tame Solomon, and wild hyenas are part of my day-to-day world. Never before I set foot on this continent did I expect that the animals I am working with would be involved in something supernatural, and at times really evil. To me, hyenas are no more supernatural than any other creature, and I have my own relationship with them. They are difficult, destructive sometimes, but they are not evil. For the people of Africa, however, things are different. I am thinking of this again in the place where I am right at the moment, not far from the national park's boundary, remembering what happened over the last few days.

Sitting in my stationary Land Rover, I am surrounded by yobs, by ruffians. There are four of them, who appear less interested in me than in the vehicle, banging against it, assaulting the tyres. Had they been people, I would have been seriously concerned.

This lot, however, I rather enjoy: they are youngish, wild hyenas, good looking with their thick fur, bright spots and lovely rounded ears. It is the middle of the morning and I am parked close to their den, with another two adult hyenas asleep near the entrances, out in the middle of a small plain. One of them is an old fat female, one of those Amazons suckling a couple of cubs: I know her well.

The car is shaking, as one of the boisterous bunch is biting a front tyre. I am not concerned: it has happened before, and they do not bite through the rubber. Then a heart-rending, crunching noise, and looking in the side mirror I see one of the thugs biting my rear light, splintering it. That is enough: I bang on the side of the vehicle, everybody jumps well back, one of them yells. Once again it reminds me how human they seem to be, these hyenas, how much like a gang of destructive teenagers on the rampage in town. No other animal has ever struck me like that. I think this nigh human streak is one of the reasons why hyenas feature so prominently and uniquely in African witchcraft.

The den that I am watching here is in the western parts of the Serengeti National Park near Kirawira, not far from Lake Victoria. Jane and I are camped nearby along the Grumeti River, far from roads, some 50 kilometres from our house in the centre of the park. For hyenas at this den it is an easy walking distance to villages outside the national park. Only a few days earlier, I had

Looking for possibilities.

seen the old hyena female I am watching now at our home in Seronera. There she was many kilometres from her den, eating from a wildebeest carcase close to our house in Seronera, and with an obviously very full belly. She is an old acquaintance, known to science with a number in my records, but we call her Dollop.

Leaving cubs without food for days on end, the Serengeti hyenas commute over enormous distances to wherever the wildebeest migration is. Near our house, Dollop was surrounded by the huge black herds of wildebeest, of many thousands of animals. Hyenas were present in numbers, too, and afterwards I see the large, fat females such as Dollop returning back to their dens to their cubs, with their legs literally chafing against a large udder full of milk, having been away from their cubs for days. Walking steadily, quietly and silently, walking, walking without stopping till they are home again, many kilometres away.

And quietly is how a hyena walks, not only to where the wildebeest migration happens to be, but also to human settlement outside of the park. It walks around houses and *manyattas* at night, picking up pieces of edible matter – whether they be bones, rotten bits of animal skin, or someone's shoes, a corpse, livestock or whatever. It comes silently out of nowhere – until several hyenas contest each other's right over some delectable piece, when hair-raising noises ensue. Their quiet appearance and disappearance is frightening to people, especially because the hyenas' sounds are so unreal, at the same time so human.

It is indeed weird for anyone to hear very high-pitched screams and yells coming out of the large, rounded body of a large hyena. Their loud, staccato 'giggles' are like those from a mad person, and they are mixed with deep growls and howls, all together the cacophony of a very aggressive orgy. It seems a long way from their single, long-distance 'whoo-oop' calls, which for me are the sound of Africa. I found their giggle to be their sound of fright, the call of a hyena being attacked by a growling other or by a lion, and when that giggle gets more intensive, it becomes a yell. It is one piece of their communication, the elaborate system of calls within the clan. It is intelligible animal behaviour.

But for a frightened human in the dark, many such sounds together are hauntingly human, supernatural, a witches' Sabbath. They are the calls from creatures that move noiselessly over enormous distances, sounds that are quiet, then suddenly overpowering.

Camping out here in that extreme western part of the Serengeti National Park, I need to get spares for the Land Rover. The nearest place is Musoma, a town on the shore of Lake Victoria quite a long distance away, and one early morning when I drive to Musoma from the Serengeti National Park, on reaching the suburban shanties I notice a dead hyena along the dusty road. Obviously it is a victim of the mad African road traffic at night. It appears unscathed, a young adult male.

When I pass it again on my way back, a couple of hours later, the carcase is still there, but much diminished. Three men stand around, looking at it, one holding a *panga* (machete). When I stop and walk up to the party, the people drift away, and I notice that of the hyena corpse, the tail, ears, genitals, one of the legs and large sections of skin have gone, neatly cut off.

Back home in the Serengeti, I tell Stephen Makacha about it. He is my assistant in hyena research, an intelligent, well-educated and a rational man, who comes from this area near Lake Victoria. 'People use hyena parts for protection,' he tells me, but then immediately changes the subject. I smell witchcraft, the ever-present evil here in these parts, officially outlawed but a powerful force. I feel that Stephen knows. He is right, people do want hyena parts for protection, and later I get requests, several by letter or directly, for 'bits of hyena', from herders to dry and feed to their cattle, or to rub into their own skin and protect themselves. Protect against what? Against evil.

A month later, Stephen wants to take a couple of weeks off to go back to his village. My reply is 'Fine, but when you are there, do talk to your people about hyenas, and let me know what is happening.' On his return, Stephen is very reluctant to come back to the issue, and it takes me weeks to drag some detail out of him.

What I get, told hesitantly and with some embarrassment, is 'I have been very frightened', and a promise that he will tell me later. Finally I get his story. I learn that one evening on his leave he had gone up a hill with a group of others. In the dark, confusion reigned, and he saw several people come down the hill, naked and riding hyenas. He himself was knocked down by one. There were flaming torches, great excitement and fear. I mumble objections, such as 'People are too big to ride a hyena.' But Stephen insists, and tells me with some finality, 'I really do not want to talk about it.' And that is it.

Over the years, I talk to many Tanzanians, people out in the bush and village people, politicians, even academics. I talk about my interests in animal behaviour, and about my research on hyenas. Almost invariably, people laugh: 'Hyenas??' I am known in Swahili as bwana fisi, Mister Hyena, and when I mention this, people laugh. Some of my colleagues are known as bwana tembo (elephant), bwana simba (lion) or bwana punda melea (zebra), but they do not get that nervous laugh. My animal is known to be something evil, weird.

Witches in Africa are people, usually men, living anywhere in any of the tiny villages. They are not recognizable as such because witchcraft is illegal, and it is difficult to get people to talk about it. I am told that witches keep hyenas, in chains, but I have never seen this. I am told that witches ride hyenas at night, and that this explains why hyenas have sloping backs. I am told that witches milk their hyenas, hyena milk having magic powers. Witches make butter out of hyena milk, and they fuel their torches with this butter. Of this there is proof, people say: when you walk in the bush in the morning, you can often smell where witches have passed while riding their hyenas, from the clear acrid smell of burning butter that dripped from their flaming torches.

I am reminded of these tales when I am watching one of my favourite hyena females, Ute in Ngorongoro, walking some distance ahead of my vehicle. Steady she goes, her head close to the ground across the open grassland plains. Coming across a patch with longer grass-stalks, she stops, sniffs the stalks, then slowly walks on a few paces, slightly crouching, dragging the grass-stalks underneath her body, sliding them under her tail. Pushing out her intestine just a few centimetres, this exposes the openings of her anal glands, which leave a small blob of secretion, a yellowish paste, on the grass-stalk. When the stalk bounces

Presence around the *manyatta*. (Ineke Kruuk)

back after she has passed, it shows the tiny white-yellow blob neatly at nose level for a hyena. And it smells – of burning or rancid butter. Even from my car window, I pick up that smell of burning butter. There goes the witch, I think – fascinating!

Apart from this scent-marking, there are so many other aspects of the life of hyenas that make them paragons of witchcraft. There are the unearthly, madmen's noises when hyenas are squabbling over food, their silent nocturnal presence outside human habitation, and their long-distance, eerie and slinking walk. There is the phenomenon of the weird female genitalia, which are almost exact copies of those of males (with a clitoris the exact shape and size of a penis, and sham testicles). Hence the belief that hyenas are hermaphrodites. The biological reason for this strange structure of her genitalia is rather unclear, but perhaps it is somehow related to the female's dominance in society. The genitals of either sex are of great importance during their meeting ceremonies, and hyenas spend a long time investigating each others', both males and females showing large erections.

The witchcraft rumours may also reflect the danger that hyenas pose to people at night, with numerous instances of attacks on sleepers or people going about their business in the bush. The East African papers regularly carry news of hyena atrocities: there, just by chance I come across the case of school-master Nyerendas Luggage in Malawi, who was torn off his bike by a pack of hyenas and did not survive, and of Godano Wario, a young girl in northern Kenya asleep in her hut, who ended up in hospital after a ferocious hyena attack on her face. There is the case of a 70-year-old man in Shinyanga, who discharged himself from hospital in the evening, and ended up as hyena food. Examples are numerous, and the role of hyenas is often remarkably similar to that of wolves in Europe, which even today also pose a danger by killing people in countries such as Belarus and Russia. As they did in Holland, until the nineteenth century.

Hyenas are known to interfere in graveyards. This may go back to days not so long gone, when some tribes of East Africa's rural areas left their dead out in the bush for animals to eat. Some still do, even today. Those human bodies were consumed mostly by hyenas, witness the human hair (and Masai beads) I found in hyena faeces in the Serengeti. To some people it implies that if you see a hyena, you may be watching the spirit of one of your forebears.

Generally, hyenas are feared, and hated here. Ecologically speaking I see no surprises in that, what with witchcraft, their damage to livestock, and the direct dangers they pose to people's lives. What I find more difficult to fit in with all this, though, is the role hyenas play in folk stories, in the tales people tell their children and each other about animals. These are fables like we had in Europe in the Middle Ages – such as the stories of Reynard the fox and Grimbert the badger, and of King Noble the lion. In African animal fables the hyena invariably is the butt of the tale, the stupid animal which gets its comeuppance, which tries to be clever but has no chance against real intelligence, such as that of the hare. This hyena role does not square with the animal I know, nor the animal that people are so afraid of. In European medieval fables, it was Brun the bear who played this part.

In Africa, witchcraft is universally recognized as an enormous force of malevolence, and most or all countries have specific laws to proscribe it. People deeply believe in it, and that gives it such power. In the Serengeti, the old cook of one of my colleagues, a man in good health, tells his employer that he is bewitched by an enemy, and tells him that he is going to die. A hyena walked around his house at night, several times. Of course the scientist shrugs it off and does not believe such things, but the old man, the cook whom I know as a very friendly soul, takes to his house and three days later he dies. He believed in the power of evil, and it got him. I have to think hard and swallow when I hear this.

In local villages outside the national park, the presence of witches is denied, of course. But every so often one notices that people do know about them, and they know that witches keep hyenas captive as familiars. Hyenas are used to cast spells, I am told, and they are instructed to visit potential victims. Believe in witchcraft, distrust a hyena.

Once during our time in the Serengeti with some atrocious weather causing widespread floods, a National Parks' Land Rover is swept away when fording a river close to our house. The car is found later but the Tanzanian driver is lost, and in the village the garage staff whisper that he was bewitched. That night, in the pitch dark and in the rain, I go and help his mates and we spread out to look downstream, carrying electric torches, to try to find him. On my own, and in the beam of my torch across the river, I see a hyena walking away. I shiver. The driver's body is found next day.

When witchcraft is at large, death is rarely far away. Of course, we all know that witchcraft is a nonsense explanation. But there is one phenomenon associated with carnivores that seems almost unnaturally lethal, and its circumstances are so apparently non-adaptive that I can imagine local people being sorely tempted to use non-biological explanations. I meet it on several occasions, in Africa as well as in Europe. But only in Africa does one attach supernatural significance. It is animal mass murder.

I have never been to a battlefield or anywhere else where I would be faced with large-scale loss of human life, nor would I want to go to such a place. But I have seen atrocities on a very large scale among wild animals – and that sheer waste of life still haunts and appals me. The first such occasion is in my student days in the early 1960s in Britain. It deeply affected me, and it has a perfectly rational explanation.

It was in May along the Irish Sea, in my study area near Ravenglass, a large area of dunes, slopes of nothing but gale-blown sand, surrounded by other slopes of sand and tall marram grass. I am camping and, flaked out one night after a hard day's work, I lie in my tent, listening to the rain hammering on the roof, a gale shaking the canvas. Outside it is pitch dark; there would be a new moon if the clouds and pelting rain did not obscure absolutely everything and anything. It is one of the darkest nights imaginable, the sky is very low, it is an unusually dreadful time by any standards. It is the kind of night when, at other times, I would go out and catch rabbits for the pot by the light of a small torch, as the animals are totally confused then by even the smallest beam. Birds sit tightly glued to their nests.

I drift off early into a deep sleep, to wake up at daybreak, five-ish, to a sky washed clear of all cloud, to dunes and sands like clean slates. Fresh out of my sleeping bag, with a lump of bread in my pocket for breakfast, I head for the vast gullery, the sandy dune valleys and slopes covered by birds.

A breeding colony of black-headed gulls is almost like a single organism, with moods, with behaviour responses to challenges. Like an individual animal, a colony thrives or it suffers. Being in or near a colony I am very much aware of that, of its exuberance or of its misery. But rarely am I affected by it so much as this morning.

When I crest a dune nearest to the nesting birds, I notice immediately that the gulls are jittery, unusually nervous, panicky and traumatized, with small flocks of highly alarmed individuals hovering here and there. An unusually huge mass of them immediately gets airborne as soon as I show myself, with waves of wings swishing silently through the air. There is the occasional alarm of soft 'kek' calls breaking the deadly silence, quite unlike the gulls' usual aggressive screeching. It is weird, and it really feels like a community responding to a major, life-threatening event.

With all the birds flying around, I notice a dead gull in the nearest nesting area spreadeagled over its nest. Further away there are more corpses, scattered randomly, still bedraggled from last night's rain. Getting further into the colony, there are waves of panicking gulls above me, swooping backwards and forwards.

I am used to finding the odd dead gull in the morning, but today the number of corpses is staggering. It does not take me long to find the cause of this disaster: despite the heavy rain in the night, the sands around the bodies still show clear evidence, clear tracks of the murderers – foxes.

Moving swiftly so as not to disturb the birds any more than necessary, I inspect and collect the corpses, to do some post-mortems in camp (and also afterwards to casserole gulls' breasts: they are excellent eating, rather like wood pigeon). That morning, I bag a total of 230 corpses, having to leave a number of wounded birds so as not to cause even more disturbance. In a colony of some 8,000 breeding pairs, this number of deaths seems insignificant, but seen as a

Mass kill of gulls and terns, by fox.

heap of discarded lives, I am horrified. And what I am seeing here is only the conspicuously white adult gulls that I found killed. In addition, many chicks and eggs may also have been murdered, taken away or hidden.

From the blurred fox tracks in the sands I can reconstruct what happened during that disastrous black night of horrors. Several foxes roamed through the colony, I think four different ones. They went seemingly randomly, walking up to sitting gulls, approaching usually from downwind from a distance of just one or two metres, grabbing their victims anywhere on the body and dropping them again, dead or badly injured. Occasionally a fox would eat part of the gull on the spot. In some places I find a dozen or more corpses within a few metres.

Occasionally a fox has buried or part-buried a gull, with its classical behaviour of digging a small hole next to a bunch of marram grass, dropping the prey in it and sweeping sand over the cache with its snout. The caches that I find usually have some feathers sticking out of the ground, so they are easy to spot.

That year, over one entire breeding season, I find 825 adult gulls killed by foxes, and within the boundaries of that same gull colony they similarly dispatched 60 sandwich terns on their nests. Apart from being overwhelmed by such numbers, it is difficult not to feel a certain bitterness when recording such slaughters. It is the sheer waste of it all. At other times, when seeing a kill I always remember that a predator has to live as well, that one animal's demise is another's dinner. But that argument does not apply any more here in the colony in the dunes when foxes are wasting their own, precious resources. What is the point?

The next night, a dark one though not as bad as the previous night, I am staying in the colony to try to see what happens if the fox comes back. I am inside my canvas cubicle observation hide, horribly cramped, separated from the gulls by only a few feet, and in daytime I would have views right across a small dune valley with thousands of nests. There are birds crooning on top of my hide, there is coming and going to and from all the nests here. But come darkness, things settle down, birds sit tight on their eggs or small chicks, with only a steady background gurgling of gull calls. It is amazingly quiet, yet somehow still rather noisy. I can just see a horizon. Close to midnight, I am beginning to flag, drowsiness overtakes me and I am very close to nodding off.

Suddenly, everything goes absolutely dead quiet. No gurgling, no calls, not even alarm calls, just an almost supernatural deadly silence, really eerie in such a huge crowd of usually noisy gulls. The silence is much more effective than an alarm call, and although I can see very little in the darkness, the birds I can pick out near the hide are sitting wide-eyed with their necks up, in what I know as an extreme fear position.

Seconds later, they are all up in the air, wings swishing past the hide, and only then do I hear a few quiet alarm calls, 'kek, kek'. Some distance away there is a slight commotion, with concentrated keks. It takes many minutes for the birds to return to their nests, fluttering noisily and obviously very carefully, settling down as if nothing happened. I cannot think how they manage to land again safely, in that really pitch-dark night. Half an hour afterwards there is a similar disturbance, and several more later that night.

This is the sum total of my observations that night. When I stumble out of the hide at first light, I find another dozen dead black-headed gulls, on or next to their nests, with fox tracks telling the tales. Foxes had the entire gull population entirely to themselves, to kill whatever they felt like killing without much effort. In the pitch dark, the gulls' usually quite effective anti-predator responses are almost totally absent.

Later, on a similarly black night, I act as a four-footed predator myself in the gull colony. It is a distinctly uncomfortable exercise in the damp darkness, of which the most striking result is that, just as I thought, gulls appear to be completely losing the plot in the pitch dark. In such complete darkness they appear to lose the ability to flee, they just sit on their nests or wherever, and even when I lift them up and put them on my hand, they are extremely reluctant to fly. 'Normal' anti-predator behaviour has gone, and that is how the foxes can make what appears an unusual mistake, killing without consuming. Usually, a kill is eaten by the predator, satiating it, and this satiation stops further hunting behaviour and further killing. It stops prey being wasted.

All that is in Europe, in a country with good, rational explanations. Now fast-forward to Africa, several years on from those gull horrors. In our small wooden house in the Serengeti I go through the violence of another black night. It is pitch black, three days after a new moon, and the house rocks under howling gales and pelting rain as it can pelt only in the tropics. At least this time I am not under canvas but comfortably at home – though the animals out on the plains must be shaken to the core.

The next morning, I am off in my Land Rover, cross-country over the endless Serengeti grasslands. The rain has stopped, the world appears drowsy after its nocturnal hammering, antelopes are standing bewildered, or they are beginning to graze. After a few miles I notice that in the distance, not far from the woodlands, vultures are circling, flapping slowly; some are landing. Vultures tell me where the action is, or was, so I know where to go.

When I get there, the scene that meets me is rather confused. Usually, when I am alerted by circling vultures I find a kill, often with predators nearby, and the birds are waiting for the killers to go. This time the action is far less concentrated, and there seem to be scattered vultures – aloft or landed on the ground – almost everywhere I look. I see a dead Thomson's gazelle, apparently undamaged, no vultures nearby; they are further along. There is another dead tommy a couple of hundred metres away, then another, and some vultures that have recently eaten are sitting around, their crops bulging with food. Rather bewildered I stop and glance around through my binoculars. I am beginning to see dead gazelle everywhere, dozens of them. One animal staggers around in circles, another lifts its head but does not move away. Again, I am facing a very eerie, chilly sight in a sodden landscape, a battlefield, without any predators.

My first thought is that these animals had been killed by lightning, of which there had been a lot that previous night. But they are too far apart, and then my Ravenglass experience comes back to me. After a close look at a couple of the carcases I realize that something unusual and sinister has happened. Injuries, and a quick bit of simple dissection shows clear bite marks under the skin; the victims have haemorrhages, some have broken legs. There is little doubt that large predators have been involved, and tracks in the mud, nearby and even underneath several of the carcases, leave little doubt that it was hyenas.

All alert now, I jump back into my vehicle and start driving around, scanning the area again and again. And soon enough, I find the culprits: spotted hyenas, bloated, some still eating from a gazelle. Altogether I meet 19 hyenas in the area, a section of grassland plain some 3 kilometres across. Some are on their own, some in small groups, all are bloated with food and bloody-mouthed, each having eaten at least one-quarter of their own body weight; most of them are reclining under small trees.

There has been a mass kill that night, by the most common large predator in the Serengeti. That morning, I tally 82 gazelle dead, 27 badly injured, and there must be many that escape my count. Fewer than 20 of these victims have been partly eaten by the hyenas. The entire scene is strongly reminiscent of that night with the gulls in the Ravenglass colony, when foxes were the culprits.

Again, I am horrified by all the slaughter around me. It is not so much the fact that so many animals have been killed, but it is the horror of the sheer numbers dead, of the waste of it all, and of the inexplicability. Almost everything I see around me here, in the vast Serengeti with its millions of animals, is efficient, and natural selection has seen to that. But the ghastly battlefield this early morning just does not fit. Why should hyenas kill all these prey without eating them, and just leave them for the vultures? Why should they deprive themselves of future supplies of tommies, which are one of their common prey species, usually killed after an energetic chase over distances of up to several kilometres? For a local here, someone brought up in one of the villages nearby, the likely explanation is witchcraft.

I know that this is nonsense. Yet, my observation of all these dead gazelle does not fit our rational picture of life in the wild. There has been a big, ghastly

Killed Thomson's gazelle, just a few of the remains
collected after a massacre by hyenas.

mistake somewhere. Perhaps the answer lies in the rarity of complete nocturnal blackouts such as the ones I experienced here last night, and in Europe in the gull colony. Perhaps the gazelle have not evolved a behaviour in response to those conditions. In my years in the Serengeti I never see a sight like this massacre again, although I witness many kills, by many different predators.

Potential prey animals in the wild may be beautifully adapted to cope with hazards such as predation, in simple one-to-one encounters with their enemies. But during a pitch-black night, with howling winds and rain – it could well be lethal for a gull to take to the air in such conditions, or for a gazelle to start running. The predators, the foxes or hyenas, have found this Achilles heel. But those same predators also have not found a way to utilize such bounty, and they do not seem to be doing themselves any favours with the surplus kill.

Most people react strongly to such waste, but is that because this means a huge loss of life, or because it is waste? I dislike taking animals myself, although I do occasionally kill for the pot, or I trap rats in the house. But I do watch predation with excitement, and I admire predators with their beautifully efficient mechanism of overcoming formidable defence and protection, even if that goes wrong every so often. And wrong it does go, sometimes resulting in the greatest chaos.

For instance, wrong it goes during the time when I am following a pack of 11 African wild dogs in my Land Rover, early on a bright, sunny morning. Lithe, patchy colours, large ears, trotting elegantly and close together across the grasslands. There are not many of them in the Serengeti. About half an hour after I find them on the Serengeti plains, just after sunrise at six, they begin to chase a male tommy, catching and killing it quickly. They do not appear to be very hungry, and several of the dogs just stand around without eating anything. It takes the pack more than an hour to finish the edible remains, an unusually long time. Yet, more than an hour later, they chase another tommy for some 300 metres before giving up. The dogs walk on, in single file.

Then, totally unexpectedly, a warthog pops out of a large hole, only ten metres from the nearest dog. The hunters take up the chase immediately and catch it within a few metres, just as the confused pig turns back to return to its hole. While the dogs are biting it, a second warthog pops up from another hole nearby, and several of the hunters immediately take up the chase. But this time the quarry escapes, and it manages to reach the safety of another hole just in time, while the dogs return to finish the first kill. The warthog's noisy death throes attract a hyena from a long way away, running fast to the commotion.

The running hyena disturbs a topi, one of the large antelopes here, with its young calf. These two run in front of the hyena towards the wild dogs – and the latter abandon the pig carcase and take up the chase, killing the calf quite quickly. Now I have two dogs eating from the topi calf, two from the warthog, and the others hanging around without apparent interest in food. While they are eating or just standing about, a third warthog pops up nearby, and immediately the entire pack of dogs is after it, catching it quickly. Minutes later a fourth pig emerges, but this time the prey escapes, and this is the final act of this performance. The dogs retire, replete, to the shade of a nearby tree; one gazelle, one topi calf and two warthogs lie wasted, only partly eaten.

These kills are surplus, they are not substantially used by the hunters then or later. But the killing instinct is always active, even when the wild dogs are gorged. There must be occasions when such behaviour is highly beneficial to the hunters – though not now here, when I am watching. But at least this time the waste of food is not as horrendous as during the earlier, mass murders by foxes or hyenas.

Such observations of wasteful surplus are not unique, or even rare. Many field biologists have seen similar things, from many different carnivores. Also, every country person who keeps hens or other livestock, any farmer keeping sheep, has seen or heard stories about the wasteful misdeeds of foxes, badgers, martens, coyotes, jackals and others. Many of these are occasions when the anti-predator defence of the natural animal is somehow frustrated, in domestic animals most often by being enclosed. Rational explanations sound hollow. That, of course, in Africa is an excuse for invoking the supernatural. Local people shiver when I tell them about what happened.

In terms of animal behaviour science, I am putting my finger here on animals' imperfections which, curiously, demonstrate how the normal system works, how prey and predator are adapted to each other. In normal conditions it will be a good adaptation for a predator to grab a prey whenever the occasion arises: after all, such an occasion is a lucky break. It is only in such events that we recognize as accidents, when the system breaks down, that we can demonstrate the beauty of 'normality', of the usual evolutionary adaptation of the animal's behaviour.

Many carnivores, including hyenas, have ways to preserve food for later, when they have killed more than can be immediately consumed – it is called caching. Spotted hyenas cache food underwater, in lakes or pools. But after a mass kill, even that kind of adaptation cannot cope with the surplus.

Years have gone past since these observations, but this does not diminish my revulsion at the waste of the mass kills in the wild, of nature gone spectacularly wrong. I feel even greater revulsion of people who are guilty of such slaughter, of people shooting African wildlife for the sake of a picture, the mass shooting of animals in colonial days, and recently the killing of lions just outside a national park. Shooting prairie buffalo from American trains was civilization gone wrong, and so is pheasants being mass killed and then discarded by shooters around my present home in Scotland.

We, humans, should know better than that. For animals it is different – many of the mammals of the order of Carnivores may kill in surplus to their require-ments. But when hyenas 'over-kill', as here in the Serengeti, the power of evil gets the blame. After all, they are known to be bewitched, or at least suspected to be witches' familiars.

CHAPTER 9

Masai, people and art on the Serengeti plains

Masai greeting.

THE HYENAS' BEHAVIOUR and their hunting interests bring conflict with other carnivores and humans. They are inevitably at odds with lions, who have the same preferences for preying on large grazing animals. And there is conflict with the human species, who either hunt the same herbivores or try to protect their grazing stock from hyenas' jaws. In addition, local people also have to protect themselves. On the open plains people meet hyenas, and hyenas and people use the same rock shelters against the elements. Although I myself may have got on well with our tame hyena, such friendship between a hyena and our species is not a typical case.

Under huge, wide horizons, under blue skies with the odd white cloud, with a few rocks here and there and the odd, flat-topped acacia tree, a herd of cattle grazes. One single human figure stands next to it, dressed in just a blanket, on one leg, holding a long, gleaming spear.

Where I am, on this enormous flat Serengeti grassland, is outside the national park's borders. The plains here are the same as inside those borders before the national park was established and people had to leave. But I cannot help feeling that even inside the park, all the Serengeti is still Masai country. It is a land of nomads, people always on the move with their cattle, with their *manyattas*. It is the country also where animals are nomads, of wildebeest in herds of hundreds of thousands, of zebra, gazelle, hartebeest and eland, elephants and giraffe, and many more. On these open plains, almost everybody is nomadic at least to some extent, even lions, hyenas, leopards and cheetahs, always on the move. With this nomadism, the Serengeti must be one of the richest land ecosystems in the world.

Through my research on hyenas, even while living right in the middle of the national park where there are no cattle and no African villages, I am made aware of the presence of the Masai around us. When watching hyenas on their dens in the plains outside the park boundaries, or when following them wandering in their search for food, I notice that whenever they see herdsmen in the distance, they become alert and often flee. Serengeti hyenas are seriously afraid of the Masai herders. They are far less fearful even of me when I walk nearby, and they show no such fright of vehicles. Generally, hyenas do not like to see people walking, but their most extreme reactions are to the herdsmen with their spears. And the Masai have all reason to be armed, with all these carnivores around.

Sitting in my 'office' at home in Seronera, where I have made a small laboratory in a thatched hut under a large acacia tree, I am staring through a microscope. I need to find out exactly what hyenas are eating, so after collecting their bright white, dry faeces I pulverize them in an electric coffee-bean grinder, take out the animal hair and identify that under magnification.

Initially I am shocked when, every so often, I easily recognize the contents of the faeces as African human hair, short, black and curly. Also in my analysis, now and then I find the coloured beads that the Masai women use for beautification. Local people tell me that hyena scats are so bright white because these animals eat ashes from fires near the *manyattas*. That, of course, is nonsense: the white material is pure calcium, product of the hyenas' habit of crunching and eating bones.

With the finds in the hyena scats, I am reminded of the custom of the Masai who live out in these wilds, of leaving the dead bodies of their people out in the bush. They let wild animals dispose of them. A well-educated Masai of the National Parks' staff, Philip ole Sayelel, tells me that usually hyenas will eat a body within a day, but if they do not it will then be rubbed with animal fat and other smelly bits to encourage them. If hyenas will still not take the body, there must be something wrong with the corpse, leaving its relatives seriously worried. Philip also passed on to me the Masai belief that hyenas sometimes chase leopards into their *manyattas*, inside the thorn fence, so the leopard can kill a goat or sheep there and carry it away. Hyenas then chase the leopard and steal the goat. One never knows, but there may be a grain of truth in that belief.

Jane and I are camped between some acacias just above a salt lake on the Serengeti plan, above Ndutu. We feel alone with our tent and our vehicle: the world is ours, there are no roads, no tourists, we have a view over the salt flats with thousands of pink flamingos, and around us in the trees birds are calling in profusion. I spent last night following hyenas in my Land Rover, and now this morning we enjoy a leisurely breakfast in front of the tent. Half an hour earlier from here I had seen a pack of wild dogs crossing the white salt-flat expanse near the water.

Totally unexpectedly, a single Masai warrior walks up, his face beautifully painted in ochre and his hair plaited. He appears just out of nowhere, plants his spear in the entrance of the tent, and stands on one leg, chewing a piece of grass. He stares at us, stares at the tent, walks in and out again, and spits on the ground without saying a word.

Then he produces a few words in Masai, which neither of us understands. We try some Swahili on him, supposedly the lingua franca here, but without success. Jane gives him a large biscuit, which he eats, then holds out his hand for more. Finally, out comes the Swahili word that is almost sacred in this country: *maji*. It means 'water'. Of course, we immediately provide a mug from our drinking supply, and that breaks the ice: he produces a smile. He puts the empty mug down and, after an hour, strides off, disappearing into the bush. We think that we are alone here – but there are people everywhere, unnoticed, and they always know what we are doing.

Weeks later we are staying deep down in the Ngorongoro Crater, in the small stone house that was built by the first Europeans to come here, two German brothers who came to farm. It is a small house under a large wild fig tree, surrounded by yellow acacias, the fever trees. Apart from the resident Masai, who then still live in the crater and graze their cattle, we are the only people around, and we love to come here, with Solomon, our tame hyena.

Sitting outside the cabin along the bubbling stream with fresh water is a delight. We love the sounds, the hyrax in the fig tree, the hippo roaring in the distance, the many birds. But before long our presence is shared with three other people, Masai women, beautifully turned out with masses of coloured beads, smiling broadly, and one carrying a baby on her back. Clearly they need something, and we soon understand what it is: medication. The baby is not well, and the women are pleased to get some aspirin, which is the only thing we have and hope it will help.

But one of them has something more complicated. She shows us her ear. Masai people habitually puncture large holes in their ear lobes, and stretch the lobes as far as they can so all kinds of decoration can be inserted or hung from their ears. In her case, though, this has gone too far and the long, thin ear lobe is broken, hanging down and bleeding. Jane manages to skilfully reattach the bit with plaster and disinfecting cream.

That is in Ngorongoro, but also just about anywhere in the Serengeti I am reminded of the presence of the Masai. It is their country. Apart from the Masai, there are several other peoples as well. The Serengeti National Park area is part of Masai land, but on one side of the park live other tribes. Among them, the Wasukuma are also cattle herders, though they are less nomadic than

Masai *moran* (warrior).

the Masai, and these two peoples are often on a war footing. One day I meet their warriors driving stolen cattle through the national park, with some of the animals at the back of the herd dragging trees behind them to cover their tracks.

That evening in the national park village, Seronera, I sit on the veranda at night, listening. I often do this, to hear the hyenas call with their magnificent 'whoo-oop', or a lion roar, to be alerted to hyenas having killed prey or clashing with lions, or just to hear the cicadas. Tonight as on the previous two nights, I hear the African drums in the village, endlessly and relentlessly, as a rather unsettling presence. Police have brought some 200 cattle to the village with a dozen Wasukuma warriors, who were caught after a battle with the Masai right inside the national park, in which two of them were killed. Somehow, the police will sort it out, but in the meantime the drums continue throughout the night. It is a worrying sound, as presumably it is meant to be.

In our early days here it is my tame friend Solomon who draws my attention to Serengeti's *kopjes*. The piles and outcrops of huge rocks and granite boulders form caves and sheltered overhangs, highly attractive to animals and people. Hyenas are reported to be cave dwellers, so when Solomon is missing during one of his escapades, I climb through a large *kopje* close to our house, calling him, and he emerges from a dark corner, begging for food.

With a great deal of animal life in these places, there often are hyena faeces; leopards also love the shelters, and so do baboons, owls and many others. Remains of pottery, flints and fire sites are witness to people who lived here before the Serengeti became a national park.

That very first time I enter that *kopje* to find Solomon, I am massively intrigued to find a somewhat faded, black rock drawing on the granite boulder overhanging one of the shelters. It is primitive and simple, but clearly depicts a cow. The artwork is considerably worn, but it does not appear to be very old. It vividly reminds me of the Stone Age art I had seen and admired in caves in France, in Lascaux, Niaux, Font-de-Gaume and other sites, and in Altamira in Spain. Even during those early cave visits in my student years, I had wondered whether this kind of art might still be practised today.

Clearly it is, as the one aspect of human culture in the wilderness around where we live. Seeing it, I am reminded that here in Africa, in the Serengeti, there is no history. But there is archaeology, and there is rock art. Reading about art in the French caves, I was told that neolithic hunting people acquired magic, supernatural powers over animals by drawing them, and great significance was attached to the various abstract signs among the figures on the rock surfaces. But there was no evidence for this, and even in Europe I never felt happy with such 'explanations'. In Africa, I feel that here it is again, the supernatural. Having learned about the practice of witchcraft and magic in Tanzania, my appetite when finding rock art here is whetted.

After I return from the *kopje* back home with Solomon, I happen to meet Myles Turner at the petrol pump in Seronera village. He is one of the Serengeti park wardens who is always interested in everything. I tell him about my discovery of the rock drawing, and proudly take him over there. I may be very enthusiastic about it, but he is not impressed – 'I know some much better ones in a *kopje* about 60 miles away.' During a very dry spell in the early 1960s he was there establishing the national park boundary and, desperately short of water, had found a small well near a *kopje*. A few yards from the well was a large rock shelter in which an old Masai and his child were living, surrounded by marvellous drawings of animals and Masai shields. 'That is really art,' Myles tells me, 'your important discovery is hardly impressive.' So much for encouragement.

Weeks later Myles and I are on our way to the *kopje* that the Masai call Oldonyo Osoito Enkishu Onyoky, or in plain English, the 'Hill of Stones of the Red Cow'. Myles needs to do the park boundary again with a grader, and I join as I need to get to know the area. While Myles and his park rangers are busy, I go ahead in my Land Rover to the *kopjes* that Myles describes. Soon I am crouching in a large rock shelter under an immense boulder, looking at magnificent paintings and drawings all around. Paintings of animals, of humans and of the shields Masai use in warfare. And interestingly, it may be just inside the national park, but there is much evidence that the place is still in frequent use by Masai people – there are fresh ashes, bones and some sticks for drying meat, and in the corners are sleeping sites for people, made of hay. Also, there are quite a few very bloodthirsty fleas.

Rock painting in the Serengeti: Masai shields, people, an elephant and cattle.

I am very excited, as the paintings are not only beautiful and some highly styled, but also some are fresh, made by the people now living in this area, an enormous distance from any towns and roads. It is a fabulous discovery, and the art bears a striking likeness to paintings and drawings made ages ago in caves in other parts of the world, in Europe, South Africa and elsewhere. The Masai responsible for this must still be very much alive. A considerable amount is known about their culture, their relations with animals and other aspects, so here is a unique opportunity to find out about the cultural background of prehistoric paintings anywhere.

Under the large boulder, I am sitting in a kind of chamber, about five metres long, which people have fenced off with thick thorn bush, with a metre-high entrance. The granite is covered with about 30 drawings and paintings of animals, and a number of strange symbols, in white. Along the top of the shelter on the huge boulder is a thick, horizontal white line of animal fat, strategically placed so any rainwater running down the granite drips off, keeping the shelter dry. I clamber around, and find that the other side of the boulder has an even larger shelter, also well used, with even more drawings and paintings.

In all there are almost 200 figures on the rock, most of them of cattle, the typical Masai humpback zebu cattle. The most beautiful one is a black outline, filled in with grey, in a natural pose. That is in fact the only grey figure, the other ones are in black, either in outline or filled in, some are in ochre and there are several whites. I also find quite a few simple, rather childish ones.

I am especially interested in the depictions of wild animals, of which there are fewer. There are 70 drawings of cattle, 43 of people, 8 of lions, 3 of elephant, 3 of giraffe, 1 of wildebeest and 1 of an ostrich. The cattle are all bulls, each with an exaggerated large hump, horns and penis. There is a large number of Masai warrior-shields, with different coloured symbols and patterns.

Drawings of Masai cattle.

The wild animals are those that the Masai are especially involved with – I know that Masai do not hunt, they are herdsmen. I am disappointed not to find any pictures of hyenas in the shelters, but the other carnivore that preys on their flocks, the lion, is well represented – and always as males. Interestingly, lions in these drawings are in scenes in which they are surrounded or confronted by warriors with their spears lifted, and some are shown in an aggressive lion

Drawing of a wildebeest.

posture, with their tail up. Cattle are always shown independent of any other figures.

Elephant are important to Masai, for they are admired as being cunning and clever, they can be dangerous and they attack when encountered. Giraffe provide people with thick and strong hide to make sandals, and a tail for a fly-whisk; wildebeest are a nuisance to cattle as they clean out the grazing and they empty water holes. They also carry diseases that may be fatal to cattle. Ostrich feathers are used for ornamentation, for the men's headdresses. All these animals from the artwork have a practical place in day-to-day Masai life, positive or negative. There are no pictures of animals that are merely beautiful, such as zebra, gazelle, impala or kudu. And alas, no hyenas.

One of the black elephants in the pictures has large, white tusks, and one of them, suggestively, is shown with a batch of white spots on the location of its heart, the place to insert a spear if one were that way inclined. There are drawings of white people (white painted in animal fat), typically standing with their hands on their hips (as the white man does when supervising his labour), while black people carry spears or swords. There are simple black drawings of vehicles, including one that looks like an old Model T Ford.

Shields, an aggressive lion, men with swords, and a bull.

I spend a whole day in the shelters, taking notes and photographing. More and more I want to talk to the artists, to ask what they mean with their pictures, and to see them draw. But during this trip we do not come across any people in the area, although there are many signs of the presence of cattle. People are keeping away from us. As Myles says, it is inevitable that the Masai distrust us, because he, as park warden, often has to remove them and their herds from the national park, where they are not allowed to be.

Six months later I am back again, with Jane. In the interval I had learned more about the Masai people, talking to several of them including Solomon Ole Saibul, who would later become Director of National Parks in Tanzania. I was told about the nomadic existence of this old warrior tribe, about their cattle, their use and opinions of wild animals, and of their lion hunts. Young warriors, called *morani*, spend weeks of seclusion away from the rest of their people, away from any women, eating only meat, and they do this especially just before one of their big raids, the cattle raids on other tribes. Probably, I am told, the Hill of Stones of the Red Cow is one such place that is sometimes used for seclusion of the men, for the occasion of what in Masai is called an *olpul*.

After setting up our tent close to the site, Jane and I set out in our Land Rover to find where the nearest Masai live, to find their village, the *manyatta*. The country is covered in thick thorn bush, it is difficult to follow the narrow winding Masai footpaths, and it takes hours to finally come across a *manyatta*. It is a circle of some 20-odd low huts, made of branches covered in cow dung and mud, and surrounded by a thick fence of thorn bush. Outside the *boma* (thorn fence) are boys with small herds of calves, and under a nearby tree a group of eight elders play *bao*, the universal board game of Africa.

Initially the arrival of my Land Rover causes consternation, but once the people see Jane they are prepared to talk, and several of them can speak Swahili. Fortunately, three of the men agree to come back with us to the rock shelters, to show us. They talk with some pride, they show us how they make their drawings. Obviously, we are not talking to any outstanding artists here, but the drawings of shields, which they make for us on the rock face with charcoal and in our presence, are part of the set that is already there, as much part of it as any of the other drawings. They show me a wild dog.

The first point I learn from our three guides is that the display is part of a process of education, of older people telling younger ones what to do, and of herdsmen boasting to each other about their cattle, especially about their best bulls. The drawings exaggerate the large horns, the penis and the condition of the big zebu hump. No nonsense here about acquiring magical powers, or about witchcraft: art is just an aid to communication about day-to-day life.

We look again at the many pictures of shields, each with a different design on it. I am told that the young men are divided in age-classes, each class with its own shield design. The pictures show warriors with shields as well as spears and swords, and in action in their main feat of manhood, the killing of a lion. When our guide is talking, I can picture the scene inside the shelter with the men near the fire, with their endless stories, week after week. 'The lion was aggressive, he was awesome, huge and dangerous, and we killed him with our spears like this, and this here is how we used our swords.'

Stylized bull.

During a meeting before an *olpul*, older men demonstrate with pictures how a lion is killed, either by a group of men with spears, surrounding the lion (depicted in drawings with a black circle around the lion) and spearing it from all sides; or, it is killed by a single, exceptionally brave warrior with a sword, after getting the lion to bite at his other hand in which he has a short stick, so the lion's mouth is stuck on it, wide open.

Cattle play a major part in these *olpul* sessions in the shelter, which last for days and even weeks, while the men gorge themselves on meat. As the eldest of our three guides explained, it is during these get-togethers that they paint big bulls, 'Now I will draw you one of mine.'

In later years I find more art in rock shelters throughout the Serengeti, though none as large and extensive as that found by Myles. I find another big one right in the centre of the national park, in what are known as the Moru *kopjes*, which are very large and densely wooded. The art in them is remarkably similar to that in Oldonyo Osoito Enkishu Onyoky. Elsewhere I come across a simple slab of granite next to a *kopje*, with a magnificent painting of a Masai shield in ochre, black and white, a lion in a defence posture inside the shield. It is another wonderful discovery of an imaginative picture. The drawings are unique, and in a style different from any others in Europe, or in southern Africa. I hope that this tradition among the Masai people will persist for a long time, even though they can no longer graze their cattle in the national park.

For me, art and science come together in these drawings in the intriguing *kopjes*, used by hyenas and so many other animals. The idea of 'my' animals strolling under the huge rock shelters and past the paintings is tremendously thrilling. I am also pleased that the Masai rock drawings and paintings, made by these people belonging and living here in these uniquely wild areas, have

Masai shield with defensive lion.

a simple, practical explanation and purpose. They have nothing to do with magic, with witchcraft or with religion, but in the rock shelters they are used like drawings made on a kind of blackboard, as I remember my teacher doing at school to reinforce his words. The artwork is part of men's conversation, of the demonstration of their prized cattle possessions, and of their prowess in confronting dangerous animals.

The rock paintings and drawings, amid the abundant wildlife, are testimony to what once was a marriage of people and their culture to an environment that could only be richer than it is now. How sad that pastoralists, denied access to the national park, no longer graze their flocks on the same Serengeti and Ngorongoro pastures, where gazelle jump out of the way of running hyenas, or lions roar. No longer can the Masai practise their skills as rock artists there.

CHAPTER 10

Striped without a clan

Striped hyena with tortoise.

W HEN WALKING IN the gull colony in the dunes of Ravenglass, I came to admire predators. Whether birds or mammals, they have a certain glamour about them, sleek, fast and beautiful, compared with the rabble of black-headed gulls on which they depended. In the Serengeti I found a paradise for such predatory animals. The African savannah has an almost unrivalled diversity of carnivores and raptors that makes a continent such as Europe feel poor. In the Serengeti alone I found 25 different carnivores, including various cats, dogs, mongooses and, of course, hyenas.

All of them have different ways of life, whether lion or cheetah or leopard, whether wild dog or bat-eared fox, honey badger or otter, one of the three kinds of jackal or one of the six species of mongoose here. And the differences between species are very marked in the ones that get my special interest, as they provide me with an insight into the evolution of group-living.

My hyenas are the spotted hyenas; Solomon is a spotted one. Spotted hyenas are the major force to be reckoned with for the people here in East Africa, for the Masai with their cattle, for other people where witchcraft is of such concern. But there are other kinds of hyena, ones that do not live in clans, that go their own, solitary way. Their existence emphasizes the enormous jump in evolution made by the spotted hyenas, and the impact that their sociality has in the world of African people and animals.

Bat-eared fox.

At night but with a very bright African moonlight, my old Land Rover is slowly bumping through the dry, thorny Serengeti acacia woodlands, off-road and miles away from any human habitation. I have removed the top half of the door of the vehicle, so I feel totally at one with Africa. Intently, I am trying to keep sight of the animal I am following, a striped hyena, but I need to avoid the many stones, boulders and tree-trunks. Suddenly, out of the corner of my eye I spot something else, a bat-eared fox ahead, standing still in the grass while watching my approach. It is a small, lovely animal; I can see its beautiful, large bushy tail half-curved, ears on the alert. I slow down to watch it and stop. The fox watches the Land Rover, alert.

At that very same time, my *raison d'être* here, the radio-collared striped hyena that I am following, dashes forward to the fox. It is using the diversion provided by my vehicle as the fox looks the wrong way. The hyena grabs the beautiful fox over the back, bites hard, and within seconds the movements of the victim stop. In a way, a very exciting event, especially since few people have ever watched a striped hyena make a kill at all. But then, that beautiful fox ... it hurts to see it being murdered.

To exacerbate my unease about the event, the striped hyena walks a short distance with its prey, then starts tearing it open, spending a good half-hour eating most of it. Finished, the predator carries off the remainder and pushes it deep into a small, dense bush, caching it. Another new excitement for me because I have always been interested in the storage of food, and the species of hyena I had been watching previously, the spotted one, only cached its surplus underwater, in pools or lakes.

One could say that the striped hyena is the *real* hyena – the animal of legend and horror, the scavenger, the grave digger – found in many parts of Asia and Africa. Here in the Serengeti I have spent many years studying the spotted hyena, but in all that time I have never seen a striped one before. Until just a few months ago, when this species began to show up in several places in the national park. It is an as-yet unexplained eruption of a population, and a couple of years later the striped hyenas disappear again.

I hate to admit it (because I have my loyalties), but the striped hyena is a better-looking one than the spotted species, about the same size with large pointed ears and a big mane. What fascinates me especially is the large difference between the two in their behaviour, perhaps most notably that the striped hyena is an almost exclusively very solitary animal, compared with the often hugely gregarious spotted one. The difference goes right to the heart of one of my main interests in these carnivores.

The spotted hyena is not averse to a bit of scavenging, on occasion prowling around our house, once even stealing and eating my shoes. But the striped one takes scavenging around houses to new perfection, often hanging around and taking scraps, not only meat or bones but also bread, fruit and just about anything that we eat ourselves. It is the striped one that must have given all

Striped hyena, alert.

hyenas the bad name of scavengers, even though it looks and behaves very differently from the spotted one.

Following the striped hyena at night in my vehicle, often just using moonlight to navigate, I see it zigzagging through the bush and across grasslands, sniffing out grasshoppers or beetles, every so often racing off and jumping high up in the air to snap at an insect it has disturbed from the ground, like a playing kitten does. From a few holes in the ground winged termites are spilling out for their nuptial flight, and the hyena is there for several minutes to lick them up. I am struck also by another difference from spotted hyenas, which usually follow small, well-trodden game paths with a steady, dignified gait, apparently interested only in large prey, compared with the striped hyena rummaging around everywhere.

The large flying termites, incidentally, are also excellent eating for us, and when watching the striped hyena I remembered catching them myself, with a hot frying pan in which they dropped from under the street lights of Nairobi. Fried, termites are a perfect nibble with a glass of wine.

After a sudden turn upwind, I see the striped hyena approach a thicket some 50 metres away. It walks underneath a branch where, about a metre and a half above the ground, I can just see a small, white-faced scops owl, hardly bigger than a starling, which is tearing an even smaller bird to bits. The hyena looks at it, then the tiny owl flaps away, dropping its booty – which the striped hyena then devours with a few quick bites, as another free snack, scavenged. Half an hour later it arrives at a tree with lots of fruits underneath, a *Balanites* which has dry fruits, with a large stone and little flesh, very sweet. It is occasion for a good half-hour of feasting.

Its foraging is much more varied than that of the spotted hyenas. Only days later in the early evening, I watch the 'stripey' dashing into a bush, just missing a spur fowl which explodes from it. Moving on, it comes across a family of cheetah, a mother with two small cubs. Its large mane raised, the hyena slowly approaches; the cheetah mother faces it, standing next to her little ones. The hyena makes a run for it, the mother cheetah charges at the predator and chases it away, for a good 50 metres. This repeats itself several times, the cheetah mother even biting at the hyena's hindquarters before the hyena gives up altogether.

On another day my neighbour Hugh Lamprey, who lives a couple of hundred metres from us in the bush in Seronera, spots a striped hyena slowly approaching his cat in his garden. It is the middle of the day, and he is fascinated by the beautiful, rather unusual visitor near his house. Slowly the hyena approaches the cat, step by step, hackles raised. Unfortunately (I think), it all becomes too much for Hugh, whose wife loves the cat – and he runs out shouting.

The striped hyenas I am watching in the Serengeti live in tsetse-infested woodlands (unlike the spotted ones on the open grasslands), in dens often under huge rocks, but sometimes just dug into the ground. The dens are mostly large, single holes, nothing as complicated as those of spotted hyenas, but even our one-year-old daughter Loeske finds them very attractive, crawling right into one (when I know for sure that the inhabitant is not at home). Both of us

My daughter Loeske at a striped hyena den.

get bitten by the maddening tsetse flies. The reason why I am visiting these dens is the presence of many food remains, which the striped hyenas carry home in their mouth and consume there, or they carry it for their cubs. Often they regurgitate the indigestibles (such as hair and feathers) next to the entrance, where I collect them to work out what they have been eating.

Almost unavoidably, when I am watching I am especially sensitive to differences from the striped hyenas' spotted cousin. But equally interesting are the similarities, for instance their scent-marking. Both spotted and striped hyenas drag their anal scent glands over long grass-stalks, walking half-crouched with the stalk between their hind legs and leaving on it a whitish paste.

It is night-time between the trees of Ngare Nanyuki, on the edge of the Serengeti grassland plains. The stripey is asleep in front of its den, and I sit comfortably in my Land Rover, waiting for things to happen, if they do happen at all. I listen to crickets, to the odd crowned plover, distantly a lion roars. Far away a spotted hyena calls, its melancholy 'whoo-oop, whoo-oop, whoo-oop' going on and on, but the striped one doesn't show any signs of interest.

Suddenly it jumps up, and stands with ears pointed. Slowly it walks away with hackles raised, and after about half a kilometre it comes across two spotted hyenas, one crunching a large piece of wildebeest. The striped newcomer circles the spotted ones until downwind, keeping well out of range from the eater, and stops, while one spotted hyena glares at it. And that is all: the striped hyena walks off again after about five minutes.

The two species only have a fleeting interest in each other, with spotted hyenas clearly the dominant partners. They steal food from each other when possible, but the spotted ones rule – and not just because they are larger, but also, whenever there is a spotted hyena, chances are always that there are more, and quite obviously, the striped one does not take any risks with its spotted cousins.

This is close to the nub of my interest in the two hyenas, the social distinction. One of them is solitary, but the other gregarious while also keeping open the singleton option. The solitary one feeds on small stuff, scavenges or eats fruits; the spotted, social hyena hunts mostly large mammals and only scavenges if it can. The latter's social life is crowned by a highly complicated repertoire of communication, especially with sound and with many special behaviour patterns, while the solitary striped hyena is silent and uncommunicative. I am convinced that these and many other differences can be explained by their ecology, through feeding habits.

Interestingly, there are still other members of that same hyena family. One of these is also a very solitary one that lives here in the Serengeti, but for the fourth hyena species I later have to go down south in Africa, to the Kalahari. Before I ever saw an aardwolf in the Serengeti, I lived under the apprehension that it looked like a smallish striped hyena, misled as I was by pictures in several guide books. Based on this false similarity that is only to be found in tourist guides but not in real life, scientists have speculated that the aardwolf is a mimic of the much stronger striped hyena, for protection.

But an aardwolf does not look anything like a striped hyena: it has only a few stripes, and there are other differences. It is a strange animal, with some curious and fascinating adaptations, some of which I noticed after I found a dead one on the gravelly road through the Serengeti plains. Its teeth, for instance, or rather the lack thereof: its molars are no more than a few, rudimentary small pegs. These pegs are set far apart, in huge contrast to the enormous chewing apparatus of the other hyena species. The aardwolf muzzle is black and hairless, its skull shows unusually large ear bulbs, and its tongue is strangely short and broad, almost square.

If any carnivore is an amazing contrast to the rest of its family, it is the aardwolf, a contrast in appearance, but especially in its behaviour. It is a weird species. No other animal would have found me kneeling on the Serengeti plains, in the middle of nowhere, sniffing between the grasses and listening to little whispering noises from termites. Or I might be smelling the air hanging above a large wildebeest midden. No other carnivore, for its size, produces such enormous droppings. Of course, the aardwolf (earthwolf) is no more of a wolf than I am, but if we only had its scats to go by, we would be excused for thinking of it wolfing down vast quantities of earth.

Sitting in my Land Rover, in the evening light fading out over the Serengeti grassland plains, my binoculars are trained on this fox-sized animal, quietly

Striped hyena resting.

moving between the grasses, its nose low and its large pointed ears moving around. After all my experience with hyenas I am fascinated, because this animal is so different and yet so closely related. For one thing, I notice that its manner of scent-marking grass-stalks is exactly as the spotted and striped hyenas do it.

Suddenly the aardwolf turns sharply, upwind into the slight breeze, and it walks a couple of metres. Having found what it is after, it starts licking the ground – and this is its principal way of foraging. End of story?

After it has finished with that small patch of ground half a minute later, the aardwolf walks on, but I pinpoint the place where it was, get out of my vehicle and walk to have a look. The main reason why I don't walk all the time when watching these animals is that, like almost all wildlife here, they are frightened of people. But they do not associate a car with that danger, and the Land Rover is a kind of hide. When I get to the relevant spot, the aardwolf now some metres distant, I have a really good look, on my knees on the ground, but apart from the odd termite and a few small termite holes, there is nothing but grass and earth – and I am sure that this aardwolf was not eating grass.

The next time I see it turning upwind and stopping to lick the ground, I decide that, for the sake of science, the aardwolf has to suffer a bit, and I immediately walk straight up to its patch, giving the animal a fright and it runs off. I find termites, hundreds of them, in a concentration the size of a couple of large hands. So that is the secret, and also, I notice when I stir the termites with my finger, there is a distinct smell around, of resin, the smell of fresh pine logs on a fire in Europe. A couple of minutes later, almost all the termites have disappeared down their little holes in the ground.

That smell must be the secret behind the aardwolf's upwind turn before it finds the termites, although I notice that there is also something else that could be important: there is a soft, delicate noise. A patch of termites rustles – but you've got to have a good pair of ears to pick it up. An aardwolf, of course, has just that, large pointed lugs above the big ear bulbs that are part of its skull.

The termites that are the aardwolf's delight are not just any random species, but of the many here in the Serengeti it is just one, of the family *Nasutitermes* (meaning 'termites with noses'). They defend themselves against their many enemies in a peculiar fashion, by squirting nasty chemicals. Mostly, these enemies are ants, but the aardwolf also counts, and when the worker termites are quietly chewing grass an aardwolf finds them in dense patches. When those workers are attacked, licked up by a broad tongue, a signal goes down into the termite nest which brings out the soldiers, en masse, while the surviving workers scuttle down out of harm's way. The soldier termites don't have jaws, but a snout that squirts terpenes over a considerable distance towards the enemy. That is what my nose picks up when I am crouched over a termite patch where the aardwolf was feeding.

With the termites, the aardwolf ingests a lot of earth as well. Hence the size of its droppings. I notice from these huge scats that they also eat quite a few other insects, such as ants, beetles and various other termite species. But these sink into insignificance compared with that one termite, which only has a scientific name, *Trinervitermes bettonianus*. If one were to believe in intelligent

Aardwolf foraging for termites.

design, here is an excellent example of a species almost made as the ideal prey for aardwolves. It is very common in the Serengeti, in biomass possibly outstripping all other species of animal.

Trinervitermes are not builders of large mounds, and are active at all times of the year, easily accessible on the surface, when they feed on dried grass. This is in contrast with other termites, many of which construct large mounds that are difficult to access for an animal like an aardwolf. In those mounds, termites store and cultivate food, so they don't need to forage and expose themselves at unfavourable times of year. But the *trinerves* termites are easy to find at any time, if you have the nose and ears of an aardwolf. I can see now why the aardwolf has evolved these peculiar characteristics.

If you are an aardwolf, and dependent on these insects, it would be inefficient to dine together with other aardwolves: you have to go foraging on your own. In contrast with the striped hyena it is highly specialized, but here is what these two have in common: they are intrinsically solitary. This is in striking contrast to their spotted relative, the species that has a preference for hunting large prey, and to whom companions can be so invaluable.

The aardwolf is perhaps a less exciting species than its relatives. Yet it has its moments. Such as now: I am watching one walking near a large wildebeest bull out on the Serengeti plains, which is standing alone in the centre of its territory. Enormous herds of wildebeest cows are grunting in the distance. There are other wildebeest bull territories not too far away, and for me it is the Serengeti as it should be.

A short distance away from the wildebeest bull is one of his middens, a mark of his territory where every so often he drops large quantities of faeces, in the form of pellets. While I watch and the bull is some distance, the aardwolf ambles along, making a bee-line for the midden from a long way away. It digs a small pit in the middle of the large, shallow heap of wildebeest pellets, depositing its own huge scats, and scratching the old wildebeest droppings back over it. I want to know what is going on, so once the aardwolf departs I walk up to the midden which, apart from masses of pellets from the wildebeest, also shows many large products of the aardwolf. I respectfully kneel at the edge and sniff – getting the strong, pleasant smell of terpenes again. After stirring the midden with my hand, I find many termites of other species, feasting on the contents of the midden.

Back in the Land Rover, I speculate that the aardwolf uses these wildebeest middens to concentrate its terpene-smelling faeces in one or a few places. When it is foraging, therefore, it does not get confused by the smell of its own droppings as it would if these were lying about all over the place. This way, its faeces are eliminated by other kinds of termite, which take them together with the wildebeest pellets. Of course this is total, though pleasant, speculation.

The small, inconspicuous termites, such as those that I find on the plains here, have a gigantic effect on all life in Africa, on mankind, animals and plants. The total termite population has an enormous biomass, larger than that of any of those spectacular mammals even here in the Serengeti. Termites eat just about anything and everything, and they are responsible for an almost unimaginable energy turnover. And from our aardwolf it appears that termites, apart from shaping the environment, were also important in the evolution of the hyena family. Those measly little insects have been the factor that shaped the body, behaviour, ecology and society of one of the wonderful species that I am watching.

Just from being with the three species of the hyena family here in the Serengeti I feel I also need to see the fourth and last one of that group, the brown hyena, and in the wild. I want to try to understand more of that jigsaw in evolution, to see what more there is to the relation between the behaviour and the ecology of these animals. This time, the Serengeti cannot help me, because the brown hyena lives only in southern Africa. The Kalahari calls. As luck would have it, someone is already studying brown hyenas there, Gus Mills, who later would become one of my PhD students as well as a good friend. The Kalahari provides wonderful additions to what I see in the Serengeti, with some fascinating relationships between people and animals.

CHAPTER 11

The Kalahari desert: a story of sand-dunes, people, hyenas and badgers

Kalahari sand-dunes.

S CALING HIGH DUNES in the Kalahari desert far away in South Africa, my mind goes back to my student years in Oxford, to fieldwork near Ravenglass in the north of England with its large, wind-blown sand-dunes along the coast of the Irish Sea. I remember a dewy, very early morning when I join Niko Tinbergen, who slowly follows the footprints in the sand made by a fox visiting the nearby gull colonies in the night. I stay just behind Niko. He reads the behaviour of the fox as it is written on the pristine sandy surface, walking with his finger pointing at the clear footprints. He shows me where the fox turned upwind, to sniff at a small tuft of grass, the wind direction demonstrated by the ripples in the sand. Small drops, in a sand-picture of a jet of urine, point to where the fox lifted its leg over a discarded rubber boot, betraying the animal as a male.

Somewhat further, the fox again changes direction, slows down, stops, slowly proceeds again: he is stalking something. Then the tracks show a rush with his toes wide apart, the sand scattered. The fox catches a rabbit several metres away, with every detail of the murder laid down beautifully clearly in the sand, so we can study it at great leisure. The rabbit is being carried off, one leg dragging in the sand next to the fox trail. There, in the Ravenglass dunes, my passion is born for good, old-fashioned tracking of animals across the sands with details of their behaviour. It is one of the many things I have to thank Niko for.

Now here I am, ten years later in South Africa, deep in the Kalahari desert on the border with Botswana, in the country of some of the greatest bush trackers in the world. For people here, wildlife tracking is a different game altogether from what I experience in Europe. They live in a land of dunes, dunes, dunes – ridge after ridge, sand and some dry grass, some bushes, the odd small tree, stretching mile after mile. Everything looks dangerously similar. The Kalahari is a world without end – and without water. There are a few wide valleys, with dry riverbeds, where most of the herds of animals hide – the wildebeest, springbok and gemsbok – and where also one meets a lion or cheetah. There are spotted and brown hyenas, many of the predators I also know from the Serengeti. Many different ones, but far fewer.

An incident demonstrates the uncanny abilities of one who knows his animals and their tracks, and whose life has always depended on that knowledge. I am being driven, way off-road, in a four-wheel pickup, and fast. Minutes earlier my student and host Gus Mills, the driver, had managed to dart a brown hyena, the subject of his study, with a dart gun, and the animal ran off with the dart attached. We have to follow quickly, to be with it as the drug takes effect when it collapses and becomes immobilized, so Gus can put on a radio-collar. The first dart missed, but the second hit the hyena's neck and stays firm while the hyena runs off, at full speed, out of sight.

While racing after it in the vehicle, Gus's Bushman assistant Houtop, a tracker, sits on the front wing of the truck, constantly pointing the driver to which way the hyena went. How he does it is quite beyond me: at speed, clinging on for dear life, he spots every odd irregularity in the sand, every slight rearrangement of grasses, picked out by eyes that are almost unbelievably better than mine. When later I walk next to Houtop while he is tracking on foot, he points out and explains small disturbances of the ground ahead of him, even shifted blades of grass. I can hardly see the footprints, but Houtop sees a story in each of them.

Houtop knows a name for every animal track we see, and makes careful comments about what the animal was doing at the time. It is a superb performance in terrain much more difficult than what we had in Ravenglass. Somehow, Houtop puts himself in the place of the hyena, or any other animal, and he imagines where and which way he would go if he were that animal. It would have left the Nobel Prize-winner Niko Tinbergen deep in the shade.

To us, most sand-dunes look the same, but for a Bushman there is a living world out there. It is not only Houtop's ability to see detail in front of him, but also his total familiarity with the dunes that shows him as a natural hunter. After the brown hyena is radio-collared, photographed, sampled, weighed and measured, we leave it in order to return to the site where it was darted, more than a kilometre away, where the other, valuable dart had missed its target. We shortcut through the endless dunes, in a straight line to where the dart rests in a small bush. It is quickly found by Houtop and I am utterly flabbergasted by his performance – as a naturalist, I feel very small indeed. His keen eye, his awareness and knowledge of detail in this difficult, to us often featureless environment without landmarks, are almost supernatural.

Like other !Kung Bushmen, Houtop grew up tracking and hunting animals in the desert dunes, surrounded by elders and others doing the same. Not only can he see what I cannot, he is also immensely tough, running large distances following an animal's (to me often invisible) tracks, doing this in immense heat, eating and drinking only once per day in the evening. I think of myself as a field man, but this is really something else. If only more trained ecologists or behaviourists would have such ability of observation and tenacity as Houtop, and be as tough as he is, science would benefit immensely.

Tracking 'spoor' (animal tracks) is not just a skill one can learn, it is also tremendously satisfactory, and really fun. It comes into its own especially when studying solitary animals, which are so much more secretive than groups. Gus knows one of the dens of a brown hyena, and we are driving out to it, early in the morning, keeping well out of the way so as not to alarm the animals. A female lies under a bush not far from the den, fast asleep. All is quiet, and she is not going to move for a bit. To find out what she has been up to last night, Houtop puts us on her most recent track across the dunes. I am keen to walk with him following the track, but the distances are huge, he is much fitter than I am and I would only be a hindrance, so I stay in the truck with Gus. The track meanders through the dunes, crossing high dune ridges or following long valleys. It shows the brown hyena stopping in places for a sniff, or to deposit a scent-mark, which consists of a minute double-dollop of secretion from her anal glands on a grass-stalk just as the other hyenas do it.

The track shows that the brown hyena has been inspecting a bush here, an old piece of bone there, so far nothing very exciting. Houtop walks or jogs in front of the vehicle, pointing at places showing something unusual in the hyena's behaviour, where we get out for a close look and to make notes. The spoor comes out of a dense bush, where clearly our brown hyena found a lovely piece of smelly animal skin, eating part of it. We wonder if that had been found from memory, or if the hyena had been attracted from a distance by its smell. Again, the track in the sand provides the answer – some 200 metres away a smart, sharp turn into the wind had brought the animal to the bush. It had stopped there and turned into the prevailing light breeze, then walked towards the food, having located it from that distance.

Onwards after this smelly prize, the 'brownie' track meanders through a large, almost bare sandy area, hitting a patch of juicy, green tsama melons growing on the hot sand. These fruits are also a favourite for many other animals here – gemsbok (oryx), porcupines, many small rodents – but I am surprised to see the evidence of a brown hyena appreciating them. It eats large

Kalahari cheetah, resting in a bush.

chunks of the fruits. I slice a piece off one with my pocket knife and try it. I find it rather rubbery and not particularly nice – but it does contain that most precious commodity, water.

After a day of tracking, we get home to Nossob, Gus's Kalahari headquarters. In the evening we check on the large wooden cage trap not far from his house, which Gus keeps well stocked with chunks of meat to attract the odd brown hyena that passes. The trap should provide a welcome chance to catch one and attach a radio-collar, but there has been little success so far. This evening, quite late, the trap springs a surprise on us – an irate, very noisy leopard is in it, ready to tear everything and anyone to shreds, if given the chance.

Releasing this furious animal seems tricky, as it leaves us in little doubt as to what it would do to the kind person opening the door for it. The one solution open to us is to drive one of the wheels of the pickup to block the gate of the large box-trap, then quickly open the gate and drive off with great speed. We hope that the leopard will run away and not chase the truck and jump on me, as the doorman in the back. After lifting the trapdoor, I hold tight and hope that Gus will not stall the vehicle as he roars off in the loose sand. Fortunately he doesn't and it works. The leopard growls loudly and runs, fast, in the opposite direction.

Following and reading the behaviour of the lonely brown hyena through the desert is a more peaceful and considerate occupation. It gives me a sense of its belonging here in the enormous Kalahari wilderness, with the brown hyena's wanderings a total, solitary adaptation to an efficient foraging existence. I am impressed, especially after spending so much time with spotted hyenas in the Serengeti. The spotted ones (which also occur here in the Kalahari) are animals

Brown hyenas on their den. (Gus Mills)

which, even when I watch them walking on their own, appear to be aware of the other hyenas in their clan territory, always ready to take on a joint chase of some large prey, a gazelle or a wildebeest, or to join in a noisy feast with clan mates, or a battle over clan boundaries.

A brown hyena in the Kalahari desert is like the striped one in the Serengeti: solitary, not really a group-living animal. It has magnificent long, dark fur, with striped legs, and it lives off carrion, wild fruits, even insects, birds and eggs, so it is no great hunter like the spotted. The track through the desert somehow underlines this, demonstrating the animal's successful, solitary adaptation. It also demonstrates that group-living is not the be-all and end-all for a carnivore. Each kind of social organization is part of a solution to the quest for a particular type of food, an adaptation to a specific environment. Each hyena species does things its own way – with the spotted hyena evolved to be by far the most gregarious, enabling it to tackle even the largest prey, quarries that are off the menu for the other kinds of hyena.

It is the spotted hyena that, for mankind, has become the special one, first among the supernaturals, and a simple, practical nuisance to people who are dependent on their cattle and sheep. The other, more solitary hyenas are rather harmless to us. It is sad that there are no brown hyenas in the Serengeti, as they are beautiful animals with their large mane, the best looking of all hyenas. It would be fascinating to have all the hyena species there next to each other. Likely, the brown species is absent in the Serengeti because of its close ecological resemblance to the striped ones. There would be too much ecological resemblance, too much competition for both to live side by side.

One animal that I come to know almost accidentally in Africa is fairly closely related to the badgers that I get involved with in Britain years later. The African badger species has a fascinating association with other animals, one of those strange relationships for which I love this continent. I first meet it fleetingly in the Serengeti when watching my wild hyenas, but later I get a chance to see much more of it in the Kalahari desert.

This time the animal has nothing to do with witchcraft, or magic or exploitation. It is clearly a kind of badger, good looking, small, conspicuously coloured in light grey and black. This one, the African honey badger, treats us with a wild aggression that belies its size. For its stature it is probably the most fearless and fearsome mammal in the African bush. And it is not just people who keep well out of its way – so do many of the wild animals. Only to some local birds it has a particular attraction, for different reasons.

My first encounter with honey badgers is in the Serengeti, when a small pack of spotted hyenas that I am following in my vehicle across the open grasslands appears set on mischief. There are five hyenas, they have their tails up looking fairly aggressive, they bounce forward in small runs every so often, and every few minutes I see one paste a scent-mark. Probably, they are on some territorial mission. They walk as if they own the enormous expanse of short-grass Serengeti plains in front of them – and their mood affects me, too: I am whistling, ready to embrace the endless African scenery.

Before long the hyenas' boisterous spirits receive a low blow. Advancing directly towards them I see two curious, much smaller animals, approaching at a trot. Slightly larger than a European badger, all grey and black, their skin appears to flop around them, their fur shaking with every pace as if it all hangs loose around the body. They look like little boys walking in pyjamas that are far too big for them. I am struck by their striking coloration, which makes the two

Kalahari honey badger, or *ratel*.

African honey badgers, an adult female with a somewhat smaller cub, particu-larly conspicuous as a kind of warning. The lower half is black and the upper parts are a light, silvery grey.

The five spotted hyenas stop dead in their tracks, staring at the approaching duo. Both badgers continue apace, heads up, now coming straight at the hyenas. Had they been, say, warthogs or jackals or whatever other animal of that kind of size, the hyenas would have been keen to investigate them and perhaps take a bite. But this time the hyenas are wary. Two of them do a few steps with tail up towards the honey badgers, but these two also hoist their tails and immedi-ately go over to the attack – and all five hyenas scatter quickly out of the way. Their tails come down, and they are clearly frightened. The front badger lunges and snaps at one of the hyenas, who yelps. The two parties separate, and each goes its own way.

I am excited about this as I know little about honey badgers, apart from their fearsome reputation among local people, and from African natural history books. I do know that they are very different from the badger in Europe – assertive, aggressive and always on the move. After this first time when they encounter my hyenas, I meet them only briefly a couple more times in the Serengeti, fortunately always when I am in the comfort of my Land Rover.

I am not surprised that hyenas give these animals a wide berth. On both the later occasions when I meet them, honey badgers charge my vehicle when I am driving too close. They come in with a fearsome noise, a long and loud, rattling growl, which give the honey badger its name in Afrikaans, *ratel*. Once the animal bites my left front tyre, and more or less at the same time sprays its urine in the general direction of the vehicle, a cloud of the most ghastly smell one can imagine, and my car stinks for days afterwards. At the time I think that, qua smell, an American skunk must be an amateur compared with that *ratel*. I spend many years in the Serengeti, meeting scores of different animals of which many have an uneasy reputation, but only once am I charged in my car by another animal, a rhinoceros, and with that one I manage to be out of the way before it hits me. The honey badger is in an aggressive league by itself.

One reason why it is disliked by local people is indicated by its English name, its love of honey. Driving along one of the roads through the bush in Tanzania, Jane and I stop for a couple of boys along the roadside, who are waving bottles of honey at us. Bush honey comes as fairly thin liquid here, with a smoky flavour, and the boys point to some trees in the distance where we can see their beehives. They are large hollow logs, dangling from the branches, high where they escape the depredations of honey badgers, though sometimes the badgers manage to climb high enough to get at the hives. It is the typical bush method of bee-keeping, and any hives on the ground risk being taken apart, totally destroyed by the *ratel*.

The animal is immensely powerful. On one of our trips to the south in Tanzania we spend the night in a tented camp for tourists, in Mikumi near Dar-es-Salaam. During the previous night a large fridge there had been broken open by a honey badger. Although it was caught in the act by the staff, the metal door of the fridge was literally ripped off, the metal cover torn as if it were cardboard. I suppose that the terribly destructive ability of the animal will stand it in good stead when opening bees' nests from tree-trunks.

Usually, wild African honeybees have their nests in holes in trees, or between rocks. There are people in the Serengeti park, called the Wandorobo, who are here mostly to exploit these natural bees' nests, to get at the honey so they can sell it to the Masai. The Wandorobo are secretive, they are few, they are everywhere and nowhere, often on their own and living in shelters in the *kopjes*, with their bows and arrows. The park wardens dislike them, of course, as they are not here legally. Interestingly, the method with which they find bees' nests is often the same as that used by a honey badger. Just by chance, we find out how.

Jane and I are staying at a remote site, miles away in acacia country in the south-west of the national park. While having breakfast next to our tent, I notice a smallish, starling-size brown bird quite nearby, chattering and twittering as loudly and irritatingly as one can expect from a small bird. When we just sit there and don't take much notice, the bird gets closer and closer, sounding louder and more irritated. Until it rings a bell with me: it is a honey guide.

When I get up, the bird conspicuously flutters away and twitters from another branch. I approach and it moves on, I follow it when it happens again, and again. We go on, and in the end I have covered a distance of about half a kilometre, the bird leading while I just follow. In the end the honey guide settles in a tree, twittering, and lets me come very close, next to a hole in a tree from where many bees fly backwards and forwards. African honeybees are horribly aggressive, so I keep well away – but I am deeply impressed by the honey guide's behaviour. If I had been an Ndorobo I would have cut open the bees' nest, taking the honey while leaving a piece of honeycomb with grubs for the bird. The Wandorobo say that if I neglect to leave such a reward, then next time the bird will guide me to a leopard, or to a rhino. I am still waiting for that.

The local people here, as well as several observant naturalists, have seen the honey guide perform for the *ratel* just as it does for me, with the honey badger following the bird to its sweet destination. I never have the luck to see this for myself. When it happens, I expect that the honey badger makes such a mess of a bees' nest that afterwards the honey guide bird will always find something to its taste.

Intrigued as I am by the honey badger on the Serengeti plains, I find it difficult to see any more of this animal there. There are too few to see them often enough, and they appear to be always on the move, never in the same place where I saw them before. Then, years later and having almost given up on this animal, I have the good fortune to be able to re-establish my relationship with the *ratel* in the Kalahari desert, in its extensive sand-dunes and sparse vegetation.

Out on my own, miles away from anywhere in that vast area and in a huge world of desert sand-dunes with a few bunches of grass and the odd scrub, I am in the most unlikely badger country one could think of. I have to remind myself of this new reality when I think of nocturnal exploits with badgers in British green pastures, where badgers live off earthworms, or in Italian steep lush woodlands where I see them eat olives in the olive groves. That is the Eurasian badger, but here in South Africa the honey badger lives in a different universe.

Camping out on the Kalahari sands, I am staying with Gus Mills and his wife Margie, still deeply involved in their study of brown hyenas. By pure luck, the honey badger is a common and quite prominent inhabitant of the Kalahari dunes and dry thorn bush country around us, and I cannot let that pass. Gus happens to have a small radio transmitter spare from his brown hyena work, a honey badger crosses our sandy path at midday, and a dart does the rest. Now, I can watch it much more easily.

Days later, walking fast in the heat and following a radio signal from the directional aerial, I know my honey badger is just ahead of me. I have to be careful not to get too close. I need to be unobserved, not only for the sake of scientific record, but there is also an element of self-preservation. In broad daylight, I am not far from it and, walking around in my shorts without the security of a vehicle or large trees, I am very much aware of the reputation of honey badgers to attack any larger animal by the genitals, if available. When I catch sight of it again, I see it moving at quite a speed, fortunately in the opposite direction.

Its strangely rounded head, quite different from a Eurasian badger, is turning left and right, the nose often held quite high, and every so often the *ratel* almost stops. Suddenly it changes direction and slows down again. The animal moves towards two bat-eared foxes, which happen to be lying down next to the entrance of their den about 100 metres away. I am worried that the foxes might take exception to me, a walking human, but they haven't seen me yet. I stop, and watch through binoculars. Two small and beautiful animals.

Getting closer, the honey badger starts its run at the foxes some ten metres away. There is dust and confusion; the badger chases briefly after one of the escaping foxes which runs for its life, its ears flattened. Fortunately, the bat-eared fox easily outruns the honey badger, which continues its trot through the dunes to try its luck elsewhere, its nose now close to the ground. I follow again, walking quietly and almost ignoring the radio signal, but as the animal quite often gets out of my field of view I also keep a good eye on its tracks in the sand, to see exactly what it has been doing when out of sight. And when tracking I can use some of the skills I picked up earlier from Houtop.

Right next to some dead branches of one of the very few trees here, the *ratel* digs a small hole in just a few seconds, again partly out of sight. Inspecting the still damp sand I find one single leg of a large scorpion. While I am making notes I am losing ground on the badger, which is now quite a distance ahead of me. I plod on, in somewhat of a hurry. Soon the radio signal tells me that I am close again. Carefully I approach further; there is a low acacia bush and a tree some 100 metres ahead. I am getting excited when I notice a large bird sitting on top of the tree, a beautiful, blue-grey raptor, the chanting goshawk.

The badger is deep inside the acacia bush, and I see branches moving; every so often large clouds of sand explode from it. Such bushes often have colonies of rodents, both inside and underneath, and I can only assume that some of the desert rats are meeting a rather rough end, out of sight. One quite large, light-brown rat shoots out of the bush across a patch of open sand, and in a flash the chanting goshawk dives down and sails towards it, taking the rat with one claw, flying off to eat it somewhere else. For me it is a real piece of luck, as for the first

Honey badger food, with its tracks in the Kalahari dunes.

time I am seeing a splendid piece of commensal behaviour which later appears to be quite common. To such an extent that later, if I see a chanting goshawk sitting conspicuously on top of a bush or tree, I know it is not unlikely that a honey badger is nearby. It is fascinating when animals use each other this way.

After the *ratel* leaves the acacia bush, I go and check on what has happened. The rat warren is badly damaged, and there is one small piece of rat leg on the sand. I am delighted at having seen the goshawk at work at close quarters, while I continue to plod on, watching the honey badger shaking its long fur around itself.

Over the next few weeks, I collect a small pile of observations on honey badgers in the Kalahari, as they go about their business in broad daylight. Several times they catch one of the ground-living barking geckos here, a scorpion there; I see them steal the remains of a springhare from a wildcat, even take a steenbok from a (much larger) brown hyena. This was tremendously exciting to see, demonstrating the sense of awe that honey badgers instil in the other predators as I saw earlier in the Serengeti. The footprints in the sand show that the *ratel* takes many small animals which often I cannot identify, small prey such as geckos, scorpions and many things that I would never pick up without the signature tracks. They show its success in catching snakes. But most of its food is rodents, and there are lots of different kinds of desert rat in the Kalahari.

One point that I notice is that unlike Eurasian badgers in Britain, *ratels* do not seem to have a central base, a den to where they return after every day of hunting. No, their strategy at the end of every day and before dark is to stop, look around a dune slope, and suddenly start digging a deep hole, with clouds of sand billowing up behind them. This badger is an inveterate digger, and it manages to disappear totally within minutes. Soon, I expect, it will be fast asleep deep down. Almost invariably these animals make a different hole every day – something which is possible here in these sandy dunes, but would be quite out of the question in the hard soils of England or Scotland. A new den every day is a useful adaptation to having a huge home range.

In fact, in almost everything the honey badger does, it could hardly be more different from badgers in Europe. It is mostly diurnal, active by day. It is the ultimate solitary carnivore, an existence quite likely dictated by its type of prey, which includes widely scattered nests of honeybees. It underlines the interesting evolutionary move on the path towards a gregarious existence by the Eurasian badger in Britain, which is based on rich pastures and woodlands.

In Britain people are very fond of their badgers, as gregarious neighbours in the woodlands, a sympathy hardly dented by the fact that badgers may cause damage to crops or to livestock, or they may transmit diseases to cattle such as bovine tuberculosis. Here in Africa there is no such love for the solitary, aggressive and destructive honey badgers.

For me, the mere mention of the honey badger or *ratel* evokes not just a single ecosystem, but a world of the animal itself in conjunction with the African bees' nests, with the !Kung and Wandorobo people, with the little bird that guides people as well as badgers to sweetness. It shows me the larger goshawk that catches rats, the crumbs of the rich badger's table. It is a community of trackers and users of the bush who stay at a respectful distance from the *ratel*. It is a world that differs totally from that other one in Europe, of the other kind of badger that is so deeply embedded into farmers' lives.

CHAPTER 12

Nomads of northern Kenya

Manyatta of the Rendille people near the Chalbi desert.

I N SETTLED CULTIVATION AREAS such as the country in Tanzania around the Serengeti National Park, people live inside their houses with gardens and fences, inside solid protection against lions, hyenas, jackals and other dangers. But there are places elsewhere in the wilds of Africa where life is more hazardous. My interest in people who are exposed to dangerous wild animals takes me a long distance from Seronera, to places where their struggles against predation are much more evident, in northern Kenya, close to the Ethiopian border. After having Solomon, and after seeing hyenas as animals in a society of animals, I find it an eye-opener to see the nuisance and dangers that predators such as hyenas and lions pose to African human life at its most basic. I am with the nomadic tribes in the northern deserts.

In the far, hot distance a group of scattered objects is on the move. Heat blurs the images, yet they are compelling because there is nothing else for the eye to fasten on to. It is flat, windy desert at midday, and almost unbearably hot. When I get closer in my Land Rover, I can distinguish cattle, a few camels and some people with a couple of donkeys laden with large empty plastic vessels. They are heading for water.

At the well some distance away, I find several other herds of livestock, waiting patiently while water is handed up into troughs from the darkness, from some ten metres down through a chain of men. One of the herdsmen, clad in a blanket, comes up for a chat, curious about my business in this forsaken country. In rather rusty Swahili he tells me about two hyenas killing a cow two weeks ago, near a *boma* (corral) about a day's walk from here. I will go and have a look – it is exactly why I am here.

This is the Chalbi, a Kenyan desert near Lake Turkana. It is way off the tourist trails and animals are few: it is country for nomads. People here are always on the move, with their cattle, sheep, goats and camels. They go where there is some vegetation for their livestock, and where there is water. And wherever they are, they are harassed by the few wild animals that can exist here. The people's livestock is food to lions, hyenas, jackals, wild dogs, cheetah. The few wild zebra, oryx and ostriches of the desert do not support many predators.

People in and around the Serengeti have complicated and subtle relationships with the animals from the African bush. There are mutual damages and benefits, people believe in witchcraft, and often show a deep knowledge of nature. This nomad country here in the north of Kenya, with people in the desert with livestock exposed to lions, jackals and hyenas, is something different altogether. Here, between people and animals it is warfare. It is also where mankind evolved, long ago.

Where the desert runs into Lake Turkana, in the heat of Kobi Fora on its shores, *Australopithecus* and *Homo habilis* are found as fossil humans. One walks there over a multitude of fossil remains of other mammals and crocodiles. Nomadic people come and go, as they have since time immemorial, people of the Gabra tribe, and of the Turkana, the Rendille and the Samburu.

Close to where we camp a village of Samburu people celebrates a party with a dance, people joining from all directions. Long lines of warriors, beautifully made up, dance with their spears in high, rhythmic jumps, exhaling deeply, laughing. Even here in this hostile environment, people exude humanity. But they tell us that they are very troubled by wild animals that threaten their livestock and the people themselves. In my project here I try to find out how serious this is, how much livestock is lost, whether people are attacked as often as I am told, and how much of their daily effort goes into protection against predators.

Our Western civilized images of the king of beasts, and of the so-called dirty scavenging hyenas, are decidedly not shared by the people here in the north of Kenya, by the pastoralists who go around these arid, desert lands with their herds. Attempting to estimate damage, I talk to the herders about their losses, I check the evidence of kills, and I study the predators to get the other side of the story. Here, in this project for UNESCO in the 1970s, I can use my experience with predation in the Serengeti and the Ngorongoro Crater. The Kenyan government is asking what the cost is to the people of living with wild carnivores.

Livestock herders hate them, they hate both hyenas and lions. In their eyes spotted hyenas are the embodiment of evil, doing great damage, and lions are even worse; together these two are seen as by far the most ghastly scourge of the area. In the encampments and villages, the *manyattas* of the various tribes, I sit down with the herdsmen and the elders and get their stories, while Jane and our children Loeske and Johnny talk to the African women and children. I hear detailed tales of recent attacks on their flocks, and I can follow those up in the places where they happened, by reconstructions from tracks in the sand, and from remains. I find the faeces of the predators and identify contents, and by various means I try to get an idea of the numbers of lions, hyenas and others in the area.

Sitting on the ground under an acacia with five Gabra men, one of them tells of his donkey that was killed by lions, now one week ago. Next to me sits Yussuf, who translates from the Gabra language into Swahili while I make notes. I hear the usual stories of what happened, such as of the wife who had forgotten to put the donkey in the *boma* at night, and of the two lions that killed it. The herdsman points out the place where it all happened about a kilometre away, so I can go and check later.

Otherwise, these people had experienced little problem with wildlife in the last few weeks. Prompted by barking nearby, I ask them about their dogs – how many do they have? There appear to be seven dogs in this village of five huts, and they are noisy. Over the months in the desert I have come to recognize that dogs are a great help in keeping predators away – but in this case of the donkey, they obviously were not effective.

We keep on talking about this and that. I tell them about the sheep around my own house in Scotland, they tell me about their herding, and from their stories I get an idea of the numbers of animals they have – which is something I can never ask them outright. I have learned to respect all sorts of convention in these conversations, such as taking a long time before coming to the main subject of our talking here, making long detours via the health of everyone and all my own relatives, the difficulties of our journey and the road. I learn never to enquire directly about people's wealth in livestock, and never to look them

Stalking lioness in northern Kenya, a danger to people and livestock.

in the eye when we are talking. The men are wonderfully kind, dignified and interesting.

Afterwards, I walk off into the bush with a young boy, who is showing me where the donkey met its gruesome end. The tracks in the sand are still there, well preserved despite being several days old. All animal remains have gone except for some dried stomach contents, and two small bunches of donkey hair. There are the tracks of the donkey before impact, those of a lion that could have been of a running jump, and a few imprints of jackals and hyenas, which presumably came as scavengers.

A few days after our visit to the Gabra *boma* where the donkey had been killed, I arrive at another group of Gabra huts. A donkey is standing under an acacia, with a large injury to its side and a bloody flap of skin hanging down. A young boy tells us it is the work of a lion. One of the men in the village confirms it, and seems rather uninterested in the animal, as if he had given it up. A lion had broken through the thorn fence of the *boma*, attacked the nearest donkey, its hind legs clawing away at the side while biting at the neck, but the attacker had run off when several men came out and shouted at it.

The pastoralists bring their animals inside a *boma* for protection at night, often after long trips from the grazing grounds. But after our next day there, a herding boy from neighbouring Gabra huts forgets to notice that one of his animals, a calf, strays behind the main herd when he brings in the animals at night. The following early morning shows me the tracks of one single hyena, pulling down the calf with little effort and eating a large part of its hind quarters and innards. Women from the village collect the remains, for food. Later I visit a *manyatta* of Rendille people some distance away, where just two days earlier five lions broke in and killed three cattle, and seriously wounded one of the men. I can estimate the times between the losses of animals.

I still have a warm feeling for all predators, especially of course for hyenas. But I must admit that, before I started this UNESCO project in what used to be Kenya's Northern Frontier District, the NFD of colonial times, I never realized how much damage these carnivores can inflict on livestock. I see it demonstrated dramatically in these huge wild, far-away dry-country areas. Also, my evidence shows that damage done to the people by wild carnivores is not only direct, that is by attacking and killing people and livestock. There is indirect nuisance, too, as the predators force the people to build large *bomas* for protection. The herders have to walk long distances, there and back with the animals. It involves the women cutting large quantities of thorn branches from the sparse bush in their arid environment. They have to repeat this often, because the thorn *bomas* get infested by disease-bearing ticks and frequently *bomas* have to be abandoned and replaced elsewhere.

The most obvious direct damage from the hyenas and the lions is the hazard they pose to people. In all the day-and-night efforts to protect their sheep, goats, cattle, camels and donkeys, people frequently put themselves in danger. Yet they continually lose a substantial number of animals. One might object that in the large majority of these cases animals are lost after human error, because of mistakes by the herdsmen, by oversights or by being late at night coming back. It does not change the case against the predators.

Bringing in camels for protection.

I am cooped up inside a Land Rover with my two young children. It is hot and pitch dark, quite late in the evening. We are not moving, waiting for whatever may come our way in dry bush country close to the Ethiopian border near the shores of Lake Turkana. At the moment there is nothing of the usual romantic silence of the African bush, because on top of my vehicle I have a large loudspeaker, producing a ghastly racket. I am broadcasting the amplified sounds of spotted hyenas on a kill. It produces a cacophony of the animals' squabbling with each other, of their yells and screams, their growls and roars, giggles and grunts – as varied as the noises of a human crowd would be, and for us uncomfortably loud at close quarters.

I recorded the tape in the savannahs of the Serengeti, many miles away from here, and near Lake Turkana I am hoping to attract the desert hyenas to the noise, in an attempt to estimate their numbers. Loeske and Johnny sit bolt upright, staring into the darkness hoping to see hyenas, which until now have rather eluded us. It is dark, with little moonlight, and I can only just see the shapes of nearby bushes.

Two minutes go by. The noise continues, then a shape walks within sight of our vehicle. Another one gallops into view, and both slink around the car to our great excitement. With a spotlight we get some tremendous sightings of two large spotted hyenas, they are almost within touching distance. Loeske and Johnny are awestruck, and they help me make notes of the animals' spot patterns so we can recognize them later. But suddenly, I hear their rather soft, quick 'staccato grunt', which I know as a hyena alarm call. Both animals are off. We are left with only the noise from the tape, and otherwise nothing but darkness.

While we are keeping quiet and wait, a minute later the car is almost rocked by an almighty roar, from a lion immediately behind us. A large male lion stands just a couple of metres away. All three of us hurriedly close our windows. I can almost feel the effect the lion has on the children, the experience is so big from this wonderful animal right next to us in the dark. He scared off the hyenas, and my experiment shows once more the attraction of hyena sounds for the king of beasts. He scavenges even here in the desert, just as on the savannah.

The observation goes back to what I found in Tanzania: lions are attracted to hyena calls because they rob hyenas of their kills – and they are prepared to cover a good long distance to do so. Inside the Land Rover, I can tell Loeske and Johnny about the many times I followed hyenas in the Ngorongoro Crater, watching them chase wildebeest and zebra in the very early morning, just like wolves would. After the hunt, when often 20 or more spotted hyenas are eating together, the noise from the steaming, squabbling, screaming, yelling heap of hyenas is audible miles away, when very often one or more lions come running up. They chase the hyenas away to settle down for a long, quiet meal, watched from a distance by the rightful owners of the kill. Our lion here has to be disappointed.

Through such observations in different places in the desert area, I am getting very rough and approximate estimates of predator numbers. When I look at their diet from analysis of their droppings and kills, I often find the hair of cattle and sheep. My data from all sources present figures comparable to what the local people tell me of the losses they suffered. So if people exaggerate damage, as I am expecting that they would, it is not on a large scale. I know that the predators in the nomad areas are surrounded by a few wild herbivores: there are gazelle, oryx, two species of zebra and others. But despite the presence of those, people's livestock is an important part of the predators' food, and many domestic animals are lost.

Chalbi desert, Kenya.

The figures are high. I find that up to 10% of domestic animals per year are killed by wild carnivores – including damage from not only lions and hyenas, but also from jackals and cheetah. The figures are somewhat different for the various kinds of livestock, and they vary for different tribes who live in different habitats, and construct their *bomas* in different ways. There are differences in the numbers of dogs they keep, and other customs. But the point remains the same: the nomad people in Kenya suffer considerably. Most wild carnivores are protected by law, though no doubt sometimes people take the law into their own hands.

Lions and hyenas are considered to be by far the worst pests, but jackals also take many lambs and kids from the flocks of sheep and goats. Of course it is not surprising that sheep, cattle and camels fall victim. They are within the normal range of sizes of wild prey for both lions and hyenas. When driving in the area at night, I once came across a young steer, unguarded, obviously lost from the daytime grazing herd, and if I had been a hyena the steer would not have stood a chance. If I can meet such a desirable and unprotected prey, a predator will pick one up much more easily.

One morning I am talking to George, chief in a large Rendille *manyatta*. He complains about the lack of support from the government, and the inability of the game rangers to control predators. Last year, two lions broke into the *boma*, killed a young camel and severely mauled one of the men. What can he do, if he is not allowed to kill the lions? Awkwardly, he sees me as authority. I stutter something about me pressing the game department to come and help, in the full knowledge that nothing will happen.

Violent attacks by animals, also on people, are much more common from lions than from hyenas here in this thorn bush country, although hyenas kill humans occasionally. Lions and hyenas have very similar wild prey preferences, although they catch their quarry in quite different ways. They are major competitors, directly by taking prey from each other, and indirectly by selecting similar victims.

In the wilds of the Serengeti, both hyenas and lions take mostly wildebeest, zebra and gazelle – as I found when I carefully documented their prey preferences. Lions stalk their quarry, and although several lionesses of a pride may collaborate, usually one of them jumps on the prey and kills. Spotted hyena strength often lies in their pack behaviour, running large prey down, then jointly tearing it to bits. Here in the desert either of these predators will break into a *boma* when a chance presents itself. Lions may attack herds in daytime, even when the livestock is guarded, but hyenas almost only roam around at night, and they are more likely to be the predator that takes a straggler in the evening.

Although here in the desert both predators only take few wild prey, when they do so that makes it particularly interesting. Once I find two lions close to the shore of the enormous Lake Turkana, the edge of my work area for the livestock project. They are asleep next to their latest prey, a large crocodile which they killed a few hours earlier, a monster more than two metres long, and they had also killed a young hippo. How a lion might get a grip on such an impressive crocodile or even a small hippopotamus, let alone kill it, defies understanding, and I would have loved to have seen what happened here. With such large wild prey I can see that a camel would not be an oversized titbit for them.

Livestock predation suffered by the people here in Kenya is strongly affected by their possession of dogs, or rather the lack thereof. Probably because of my European background, I associate the guarding of animals, of sheep or goats or even cattle, with dogs. Not so much in Scotland or England, where sheep are left to roam on their own. But if I go to the mountains of Italy, Turkey or Spain where there are wolves, I am bound to make the uneasy acquaintance of the guarding sheep dogs, often with their spiked collars. I will not forget the huge and nasty Anatolian ones.

Likewise, some of the 24 *manyattas* I visit here in northern Kenya have numbers of village pie dogs roaming around, barking at me. It appears to be the norm among the Gabra and the Samburu. But many villages do not have dogs, especially those from the Rendille tribe where only a quarter of the *manyattas* have any dogs at all. I very soon learn that if I visit a *manyatta* where I am not being assaulted by these canines, one without barks and flashing teeth, I am likely to be confronted with many complaints from the people about predation from wild animals on their herds. Such *manyattas* suffer demonstrably more than those with dogs. It makes sense, but why, then, does not everyone here use these pie dogs for protection?

Partly, this is mere tribal preference, and for instance the Rendille people tell me that usually the dogs have no effect on livestock loss to predation. But the Gabra people contradict that, and in more than three-quarters of their *manyattas* I find man's best friend. The Gabra complain seriously about the recent government efforts to catch and destroy all free-running dogs in the official campaign to eradicate rabies. My own observations suggest that dogs are a good thing, that *manyattas* with dogs lose fewer animals, and I try to get the authorities to change their minds over their dog eradication policies. But in the government, with their veterinarians, I am up against an immovable object.

Rabies is obviously of great concern here, and the evidence is clear. Twice I meet what is almost certainly a rabid jackal in the desert, frothing at the mouth and very listless. Once I need to drive a mother with a small, six-year-old Samburu boy to hospital in Marsabit, the centre of civilization and a day's drive away. The poor little soul has been bitten by a rabid dog a few days earlier, in his *manyatta*. Doctors take mother and son off my hands in Marsabit, where they disappear into a large, chaotic medical complex, all on their own in the crowd.

Talking all this over in the Land Rover, the children and I drive across the Chalbi desert, part of the flat country here in northern Kenya. The landscape is enormous, with mountains only in the very far distance. Around us is only dry sand, without any shrubs or large vegetation, just the odd palm tree, and there is no road or track. It is hot, and a gale-force wind blows through the open windows of the car: Loeske comments that it is like living in a hairdryer. Suddenly, we are caught by a *vili vili*, a dust devil.

It howls through the Land Rover, everything shakes, there is sand everywhere and over everything, painfully so. Worst of all, the wind picks up our map, an enormous sheet of paper which takes to the air, and this absolutely vital tool goes fly-about fast, high into the sky. I drive after the *vili vili* across the sands at full speed and, after several abortive attempts covering hundreds of metres, Johnny manages to jump out when we are ahead of it, catching the map with both arms.

At the end of our four-month project in the northern deserts of Kenya, I spend a last evening at a *manyatta* of the Rendille people, a place called Kargi, while the sun sets over the dark stony landscape. Looking down from a low hill, I can see the hundreds of camels, the cattle, goats and sheep, all milling about between the huts, inside the thorn fence. Despite the racket, despite the bellowing animals, the dust, the shouts of the people, it is a scene of utter serenity. There are no hints of animal dangers, of lions or hyenas lurking about, of anything to fear. I can see clouds of dust approaching in the distance, of late-arriving herds back from their foraging day, and of women directing camels that are dragging large thorny branches of acacia bush for the *boma*, to add to their protection. For half a kilometre around the *boma* all trees and bushes have gone, and further away they are sparse. Fencing branches have to be dragged in from far away now.

Tonight in the dark, the thorn *boma* of the Rendille herders will be tested again. Hyenas and lions will be trying it out for weak spots. It is a constant battle between the people and the predators, and only they who are living in the huts there, they who are walking their livestock over the many miles in daytime, are aware of the danger, and fight it. In these Kenyan deserts, where one is close to the birthplace of humanity, there is no talk of witchcraft, also no talk of tolerating wild animals, and one does not eat game. The nomads survive with just their livestock, with lions and hyenas an ever-present danger to both people and their animals.

Strangely, further north from these conflicts there is another scenario, in total contrast. There I find a town of mutual tolerance between hyenas and people, an urban idyll deep inside Ethiopia.

CHAPTER 13

Harar, town of the people's hyena

Wild hyenas fed by the locals.

I F THERE WERE many hyenas like Solomon, with domestic hyenas just as there are domestic dogs, and if there were an urban community where such domestic hyenas were de rigueur, Harar would be that place. But Solomon, of course, was unique.

Darkness has settled over the heat in the narrow streets in Harar. The old city is locking up for the night, few people are about, and whatever life there is, it is silent. Jane and I wander about leisurely; tall houses are towering over us, gloomy and decrepit, windows and doors mostly shut. The street we are in is next to a high, medieval city wall which completely surrounds the town, falling apart in some places. This part of Harar, in the lower regions of Ethiopia, does not have the feel of Africa. It is more like the Middle East or Europe, old and worn in its streets and buildings, with dozens of mosques and minarets, with the call of the muezzin.

In a doorway, an old woman lies asleep in her dark clothes. We watch from nearby, from the other side of the street, as one single, large and fat spotted hyena walks past. The animal ignores us, casts a brief glance at the prostrate figure, and walks on – just as if it were walking on the Serengeti plains (except that there it would have been more wary of us, as we go about on foot). Coming from the wild savannah, it makes me feel uneasy seeing a hyena in this totally different, urban environment, an environment that instinctively I myself have never trusted; I am not an urbanite.

Ten minutes later, again we meet hyenas, more in the central depths of this small town, this time two animals harassing each other over a scrap of something, a bone or some animal skin. Dogs are barking, but most of what we hear, the sounds of howls and giggles and other hyena calls, are for me the sounds of the Serengeti and Ngorongoro. The backdrop here is another world, buildings and streets and waste, the set for the urban hyena. Each night Harar is invaded by hyenas. They enter along roads, through holes in the town wall or through places where the town wall is down, and behave like stray dogs. But they are hyenas, spotted ones.

In daylight the following morning we find a large market just outside the city gates, alive with people and animals, with dogs and donkeys everywhere. Here we feel Africa, or perhaps Europe as it used to be centuries earlier. When stalls close down in the evening there are edible scraps everywhere, a big spread for the visiting hyenas, and for the town's pie dogs.

An elderly man tells us about the hyenas, explaining that they do a good job in keeping Harar clean, removing bones, bits of skin or any other animal offal that people throw away. He tells us that people are not afraid of them, they like them, that hyenas do not bother people. And he tells us that often the animals are fed by people to keep them happy. A man sits just outside the city gates in the evening: he has befriended the animals and brings them bones almost every night.

Full of curiosity, we walk over there that evening from our hotel to where we find Hassan, sitting on the ground near a faint street light, surrounded by a dozen large, adult hyenas, all lying down, several of them asleep. He is a middle-aged man, sitting on his own with the animals, with a sack next to him. Obviously the hyenas are aware of its contents: they are facing it, eyeing it. But they do not try to approach further than a metre away

Harar near the city walls.

Hassan smiles at us and puts his hand on the sack. Immediately, all the hyenas are alert. Gently he extracts a bone, and quietly passes it to the nearest animal, who takes it, equally gently, then walks a few paces to eat its treasure. Next thing, Hassan takes a bone into his own mouth, facing the hyenas – one of which walks up, puts one paw on the man's legs, and, again very gently, takes the bone out of Hassan's mouth.

He motions Jane and me to sit down next to him, and lets us feed the hyenas. The animals take food from our hands as they do from his, slowly and carefully, without snatching. He points them out and names each of them. Sadly, we have no language in common, but his affection for the hyenas needs no translation. Our thoughts are with Solomon; here in Harar it feels as if he is about to come back to us again.

I am staggered: what a world of difference between this gentle restraint here, near the city walls of Harar, and the scramble for bits of carcase by the wild hyenas that I know from the Ngorongoro Crater and Serengeti plains, with the fights, screams and manic giggles. To me, these urban scenes illustrate the hugely different habitats and ways of life, the large variation in behavioural adaptation of these animals. Elsewhere, this animal is known from the East African savannah, from the deserts of northern Kenya down to southern Africa's Kalahari, even from dense tropical forests high in Kenya's Aberdare Mountains. Obviously, without much trouble the hyena can equally adapt to a totally different, urban existence, as to other kinds of environment.

Of course it is exciting to meet hyenas here in the backstreets of Harar, to see them accepted by people, and to see them fitting in so well to their urban existence. It is occasion to remember Solomon. But at the same time I cannot suppress my unease about it. Scavenging on the streets of a town and living off handouts from people must be the bottom of any existence, and one could say that these animals are the dregs of hyena society. I see spotted hyenas as the hunters of the savannah, the upper crust of the animal community, even though their reputation among people there may be evil. Now here they are street scavengers, the lowest caste of all. I must admit to similar reservations about urbanized animals elsewhere: stray dogs in Africa, urban foxes and badgers in England, or raccoons in America.

Hassan with a friend.

Only a few months after Jane and I witnessed these scenes, I am told that Hassan was killed by his hyenas, though no one seems to know exactly what happened. So perhaps the urban hyenas are not quite as changed by human civilization as first appears. Nevertheless, hyena life in Harar city still goes on, decades later just as if nothing untoward had taken place. People are not afraid of them, they feel no danger of being attacked by hyenas. But try to tell that to the herdsmen of the Kenya deserts, or villagers in Tanzania or Malawi.

When we leave town the next day and drift outside the city walls, the urban environment disappears, and with it the colourful crowds, and the noise. Instead, we wander past poor little farms, green fields, bushy rough ground, scattered trees in the hills. It is small, poor agriculture similar to many other places in the world. This is where hyenas live in daytime, in their dens not far from the town's cauldron, their paths leading into it. Jane and I walk around in the farmland and people greet us rather shyly; small boys fling stones with their lethal slings. Weirdly to us, this is part of the Harar hyena habitat, but now, in daytime, there is little sign of the animals themselves. Yet they are there, and at night they commute downtown. The real hyena habitat here is urban – alas.

The tremendous range of environments in which spotted hyenas exist and adapt themselves into several – almost quite different – types of animal, makes me wonder whether differences in their genes are involved. Or might it be that there is nothing innate in this, that cubs just pick up the habits of their mother, and continue from there? Perhaps it is a combination of these two possibilities.

Some large predators show enormous regional differences in the danger they pose to mankind. In eastern Europe wolves were, and are, a big menace to people, especially children, frequently killing them, and there were many well-documented cases of this even in the west, in nineteenth-century Holland. But in the wilds of North America wolves do not go in for man-eating; remarkably they are harmless there to people, though not to livestock. Brown bears frequently attack and kill people in the United States and in Canada, but the same species almost never does so in Europe. Cougar (mountain lion) have killed many a walker or jogger in western Canada or in California but do not appear to do so elsewhere. In all such cases one wonders if there are genetic differences, or if perhaps they are the result of some kind of learning from each other.

Of course there must be good adaptive reasons for such variety in behaviour, for such 'cultures' in animal populations. Individuals may learn from each other, though in the case of the spotted hyenas I saw no evidence of them actually teaching their young ones in my Serengeti years. That is quite different, however, for some of the large cats, for example, or for otters.

Mother hyenas may not appear to teach youngsters, but what does happen is equally interesting: somehow, young hyenas' social status is affected by that of their mothers. Other scientists, who studied hyenas after I did, found that the cubs of the dominating large females who lead within the clans also become dominating, leading individuals. Could this be through deliberate 'teaching'? Not necessarily of course, at least not as directly as is shown by, for instance, a cheetah or an otter. Those species do really teach their young. Perhaps hyena cubs just accompany their mother, and pick up social and environmental clues from there.

Just by joining their mothers on foraging trips, cubs of any species may learn what to avoid and how to stay out of danger. In other carnivores which do have more detailed skills to stalk, catch and kill their prey, we also find specific behaviours to keep their young in business. In those cases, teaching young carnivores to forage and to hunt is something so striking and interesting, that I have to write about some of what I watched, although all of it occurs far from Harar. When it happens, it is heart-rending to see what the prey, the victim, has to suffer, but one has to remember that the carnivores, the cheetahs or otters, have to live as well.

I recall when, in the heat of the middle of the day, I am parked next to a spotted hyena den in the Serengeti. Nothing, or next to nothing, is happening. Even I am getting bored with it. Having nothing better to do and scanning the area around me, I happen to see the small head of a cheetah in my binoculars, where she sits looking out of a patch of long grass some distance away. Even from afar, a cheetah head is unmistakable, this time even more so because there are a dozen or so Thomson's gazelle nearby, staring at the predator and making it even more conspicuous. A herd of some 300 gazelle grazes further up the slope. I drive a bit closer.

Moments later the cheetah gets up, and I notice that she has three two-thirds-grown cubs beside her. The cubs stay put as the mother walks out across the short grass, roughly in the direction of the herd of tommies. She is not heading directly for them, but she angles towards them and just skirts the herd, head up and looking at the gazelle, alert. Most of the gazelle flee when she is still about 100 metres away, but some, especially the male gazelle, stay where they are, facing the cheetah.

When she is still some 50 metres from many of the tommies, suddenly a small, baby gazelle jumps up from where it was lying, and runs. In a flash the cheetah is after the fragile little thing and grabs it in a cloud of dust, about 400 metres from where I am. The cheetah mother emerges from the dust, walking slowly, the young gazelle dangling motionless by its head from her jaws.

Her cubs come running up to her, and she drops the baby tommy in front of them. One cub tries to grab it, but just then the little gazelle jumps up and runs, fast. Immediately both cubs are after it. The fawn runs in a large circle, the cubs on its heels, the mother cheetah just standing and watching. Twice one of the cubs hits the prey with a paw and rather clumsily bowls it over, but twice also the gazelle jumps up and is off again. The cheetah mother watches closely, and slowly walks in the direction of the party. When the young tommy, with the cheetah cubs on its heels, again gets near the mother cheetah, this old, experienced hunter takes a few quick steps over just a few metres and quickly puts her paw on top of the gazelle, which rolls over and lies still. It is the end of the hunt.

The cheetah mother picks up the motionless tommy, carries it off and away from the cubs over some 30 metres, followed by her offspring. She then drops the dead prey on a low termite hill, and she and her family sit around it, panting. Five minutes later they start eating, finishing their meal in a quarter of an hour. The cubs had their chase and flunked it – next time they will do a bit better.

Mother cheetahs teach their cubs.

Mothers teaching their youngsters are fascinating to watch, tough on the victim, and I find it tough as an observer, too. But the cheetahs are lucky, they get their meal – and what is more, the spotted hyenas on the nearby den do not know what they are missing. If they had seen what was happening, the cheetahs would have stood little chance of keeping their delicacy.

Hyenas never get such a detailed, tolerant kind of lesson from their mothers. Perhaps there is no need, because of their simple and often gregarious methods of hunting and killing. A successful catch is much more difficult for the solitary cheetah, as it is for all cats. And for otters.

Shetland, many years later: I am perched on a stone along the shore, an autumnal sun beaming across the sea. A distant island stands out, a fishing boat, some gulls. This time I am watching otters. One of 'my' otter females is fishing just below me, about 30 metres out. I am keeping well down against a peat-bank, not showing my profile against the sky. It is low tide, and I know that the otter I am watching has two cubs, about five months old, somewhere between the rocks in the seaweed. They are out of my sight. Up she comes again, eating one of her usual prey, a small, eel-like fish from the sea bottom, an eelpout. She wolfs it down, head pointing skywards while she effortlessly floats on the surface. She looks around, and dives again with an elegant tail flip. A good 30 seconds later she pops up on the surface in the same place, this time with a larger prey, a good-sized rockling in her mouth. She does not eat this fish but swims to the shore with it.

When she lands there, she is met by her two charges, which have obviously kept a close eye on her all the time. The two otter cubs leave little doubt about what they are after, jumping around their mother and trying to reach the fish in her mouth. But the mother has other ideas, and she walks off over the rocks to a nearby, shallow tidal rock pool while carrying the fish out of reach of the cubs. The pool is a place where, at low tide such as now, I often look for sea life, for gobies, crabs and sea anemones – the pool is no more than about ten metres long. At the edge of it, right in front of the cubs' noses, the mother releases the rockling in the water, and the fish darts off in a flash. There is no way of escape from the pool. Both cubs dive after it – the mother stays where she is.

There is a lot of splashing. Swimming and diving, back and forth, it takes the cubs almost a minute before one of them catches the fish again and hauls it out. Immediately it starts eating, not allowing its sibling a share. I realize how clumsy these young otters still are – and how difficult it is, for a land mammal such as the otter, to catch a fish. A lot needs to happen before these two are ready to forage on their own. The one who is best at the game gets the prize.

Sometimes, when the mother releases a fish in a pool like this, the cubs don't manage to catch it and she does the job herself. She may release the fish again, and again, or she may just drop it on the rocks in front of the cubs. With such scenes I realize that I am watching mothering at its best, just as with the cheetah in the Serengeti. At other times I notice that the cubs also often watch the mother fishing in the sea at close quarters – the cubs float on the surface with their heads underwater, while the adult otter is diving and fishing along the bottom just below them. Cubs may learn then, but on those occasions there is no evidence of actual teaching, no evidence that the mother changes her own fishing behaviour.

Mothers also show their care in what the otter mother brings to the cubs. Her young ones get fish to eat of a size that, on average, is considerably larger than what she eats herself. That is just what I am watching here: when she gets the solid, large rockling and takes it to the cubs, it is several times the weight of the eelpout that she ate herself just previously. It is a beautiful and immensely pleasing observation. I never saw anything equivalent when watching hyenas.

Teaching their young to hunt is one of the many little parental sacrifices that we see in carnivores, whether in the form of enabling cubs to perform, or merely allowing them to be present at the action. It is often at a cost to the adult itself, as when the clumsy cubs alert potential prey. I think that in general for hunting animals, teaching is a behaviour that enables young, prospective hunters to adapt to new prey, or to new environments. There is no such complicated behaviour in the hyenas walking the streets of Harar, they are there simply from following in their mothers' footsteps and seeing her trade, resulting in the animals' total dependence on human society. They clean the town. Harar citizens benefit from this, unlike the Kenyan nomads, who never see any benefit and nothing but grief from hyenas.

CHAPTER 14

Vultures gathering

White-backed vultures, and one hooded vulture.

T HERE IS ONE GROUP of birds that quickly captures my imagination in my first African years, birds that are almost proverbial subjects of contempt and revulsion. But given my interest in hyenas, I soon get used to looking at animals that are reviled by others.

Even before I ever set foot in Africa, I knew that hyenas and vultures are often mentioned in the same breath, as scavengers. But from the beginning vultures are part of my world, part of wild Africa, and a great help in my research. Especially in the Serengeti one just cannot escape the large birds that circle the skies, distant shapes that appear tiny but by just being there dominate the endless skies. For me, vultures also are an illustration of how incredibly diverse and complicated the Serengeti community is.

High up on the rocks of one of the *kopjes* close to where we live, Jane and I are chatting to a few friends, talking about Solomon, about wildebeest and Harar, while we are looking over the endless Serengeti landscape. There are countless animals in our view, and we have a drink in our hands. We talk about people in the village, about our research. It is a scene of tranquillity. Inevitably, my mind drifts away from the conversation. My binoculars follow my eyes around, away to some specks moving about high in the sky.

Their gentle movements do not seem related to the animals and scenery below; they are majestic in their determination of where they are and where they go, without any wing movements, without the apparent lifting of a single feather. The birds may be far away, but they are mesmerizing. In binocular close-up their huge wings are fingered, head and feet out of sight, beautiful shapes circling, and circling again. The wing-spread of these enormous vultures, even so high up, is silently impressive. They stay up there, hours on end, literally effortless.

Up in the sky the birds are a world away from their other existence, from a world with these same birds sitting on the branches of a dead tree, not far from a carcase. Dark, hunchbacked, with huge beaks and claws, their big black eyes and a look of eternal fury on their face, there they are the picture of pure evil, the image of death and revulsion. Any raptors' large wings and sharp beaks have often invoked fear and the supernatural. I remember that in Holland, the country of my youth, I was once horrified seeing a large owl nailed spread-eagled onto a barn door, farmers keeping out the devil. But vultures are fascinating, and I see their beauty.

Thinking about this while watching the circling black specks from the *kopje*, I reflect that I seem to be curiously attracted to underdogs, and to animals despised by other people. To the hyenas that are the focus of my research and which are often so loathed, and to these vultures here.

Vultures appear to be almost literally everywhere in Africa, in all landscapes: 'all jobs considered'. I see vultures in urban Harar, they are outside towns, above forests, over farmland and savannah, anywhere in the huge stretches of wild country. Many islands are without them, as the birds' dependence on thermals stops them crossing wide stretches of water, and for the same reason they do not occur in cold countries. Seeing them from the *kopje* in the African scenery, they appear as an integral part of the Serengeti, to have been made for this landscape.

Apart from their in-flight beauty or cause of revulsion, it is the natural history that attracts me. And if I am searching for a real-life example of intense scramble competition between animals, I can hardly do better than looking at vultures. There are no fewer than six, very different species of them here in the Serengeti alone. When feeding at a carcase, sometimes more than 100 literally disappear in a heaving, screaming, scratching, biting heap of all these different species, a feeding compaction that must be unique among birds. The scene would be a ridiculous caricature of Darwinian competition in nature – if it were not so real. Also, it appears in total contradiction to the glorious Serengeti surroundings.

I get involved in studying these birds, curiously not by design, but more by accident. No research proposal was ever written for my vulture project, no budget was ever drawn up, it just happens when I am working in the Serengeti. I realize that my interest in vultures starts as a consequence of my initial failure to come to grips with my study of hyenas, which is my *raison d'être* here. While struggling with my main research problems, I am seriously diverted by the conflicting signals these vultures send, and by what I see as gross, unexplained inadequacies in their behaviour. Birds just are not like that.

When I arrive to live on the fabulous Serengeti plains, I am with little knowledge of any of the animals, but with serious ambitions. I try to find

Hyenas and white-backed vultures on the Serengeti plains.

answers to questions about the most abundant large predator. I have questions about their ecology and behaviour, and its impact on the many grazing animals. Do hyenas kill or scavenge, and if they scavenge, do they deprive other predators of their food, thereby requiring these others to kill more? To get anywhere with how these hyenas are organized, I go out in my Land Rover to find them, to follow these nocturnal animals, and to see them feeding. But due to my initial glaring inexperience, working at night is hardly effective. To make a start, I need daytime contact. I need to know where the animals are, and to get at least some data on their activities in daylight. Once I achieve that, I can expect to take the project further at night.

It soon becomes clear that if only I can find kills, carcasses of wildebeest, zebra and others, I am quite likely also to meet hyenas. And there is no better way to find a kill in daytime Serengeti than by looking for vultures. Subsequently I get interested in them for their own sake. It is an almost classical manner of becoming sidetracked when one's main project is somewhat beyond one. Being sidetracked has happened to many other scientists here, and someone once described Africa as a graveyard for zoologists, with its vast potential for distraction. In my case, it needs a researcher stronger willed than me to keep my eyes on the ball, on the hyenas, and not be distracted.

A vantage point on the Serengeti plains, scanning with binoculars. I know that I may be there for minutes or hours, looking for hyenas, but I am bound to see vultures, either sooner when the large herds of wildebeest and zebra are around, or later when the huge grassland plains are empty. Specks in the sky, always on the move. Sometimes, when they are at their most conspicuous, I see many vultures circling together, flying low. That happens when they have recently flown up from the ground, abandoning whatever they were doing there. Such a low spiral of vultures is an exciting spectacle, but usually it means that most of the real action is over. There may still be the remains of a carcase underneath the flock of birds, and with luck there may be a few predators still around – hyenas, or lions, or wild dogs or whatever.

The real thrill comes from those small dots in the sky when they are coming down, fast and on a steep approach. Closing in their wings are half bent, legs outstretched like the landing gear of an aeroplane. Even a long way off I can hear the air hissing through feathers when the gigantic birds plummet down to earth, before elegantly rounding out and landing with a few graceful hops. As a fast means of getting down from their great height, nothing beats these spectacular descents, and they are obviously highly effective in getting to a kill quickly. The descending birds are like a large flag on a carcase, announcing it far and wide, even kilometres away. They bring a scientist like me to where the action is, to where a corpse is waiting, or about to be produced. My presence is immaterial to the vultures, my scavenging is academic – but I am not the only pirate, and if I can use those birds to find a carcase, so can others.

Inevitably upon one of these dramatic descents, other vultures converge from everywhere, and often also hyenas, lions, jackals and others. Every single one of those is a competitor of the bird which initially spotted the treasure. If only that first one would be a bit more circumspect, it would have a meal all to itself.

That others do take advantage from those first vultures to alight, clearly responding to the high-speed descent from many miles away, is demonstrated in a 'natural experiment' that leaves some animals looking decidedly silly. I have seen it several times, but the first occasion is still strongly grafted upon my memory. It happens far out on the Serengeti's short-grass plains, in March, the beginning of the rainy season.

I stop the car. All is bright, fertile green around me, dotted with white and yellow flowers; there are rain puddles, and scattered herds of wildebeest and zebra. The reason for the interruption of my journey is the first, large flock of small, black-and-white Abdim's storks overhead, a seasonal migration probably straight from their breeding grounds in the Sudan. They are attracted to the wet Serengeti pastures. Suddenly, down they come to wade in the puddles around me, dropping out of the sky like a mass of leaves from trees in a European autumn. Their annual influx has arrived.

My own first, brief thought at seeing all those big, dark birds plunging out of the sky more than a kilometre away is 'vultures!', before I realize my mistake. But the nice thing is that apart from myself, several real vultures, as well as

White-backed vultures chased off by a lion.

two spotted hyenas and a jackal, are also fooled. They hasten themselves to the storks, similarly from a long way away. When they get here, they stand and stare rather foolishly, before flying or sloping off again, looking ridiculous and making my day.

In the next few days many more of these Abdim's storks will arrive over the Serengeti, and very soon none of the scavengers will be fooled again. Vultures, hyenas, lions and jackals, they all learn. However, they also forget, and next year, in the next rainy season at the first arrival of the migrating Abdim's storks, I will see the same thing again, and repeatedly: all of the scavenging community habitually mistaking the first storks of the season for vultures.

If there is a carcase to be had, many vultures come bombing out of the sky very quickly, and saturate it with their presence in a flapping, noisy pile of the enormous wings, beaks and talons of the different species. I keep being absorbed by such heaving aggregations of birds, although I know that I should be studying hyenas here.

A fighting collection of vultures appears to be chaos, which seems an anomaly. Surely there must be order in it, somewhere. And of course, I find that there is, and the birds have rules. After seeing quite a few of these vulture assemblies, I realize that within that mass of six different species (as well as other birds such as marabous, kites and eagles), each of them, each species, does things its own way. Yet that still leaves ample room for fierce, often violent, competition.

A Serengeti dawn is an experience beyond all dawns. The cold, damp and slightly misty grasslands are still asleep, the sun is well below the horizon, but light rapidly spreads over the dewy stalks. A bustard calls, then a francolin, and across the river valley wildebeest begin to grunt. The early-morning light strikes the yellow bark of riverine acacias. I sit in my Land Rover where I have been for hours, cold but completely occupied by the animals.

During the night that is just fading, and by homing in on loud whoops and giggles, I had found a dead zebra being consumed by 12 large, spotted hyenas. It was still steaming hot when I came to it, so the hyenas had probably killed it themselves. I saw many things happen at that kill, but the upshot was that the hyenas had been chased away by one single lion. The king of beasts assumed his role of scavenger, chased the hyenas, ate, then settled down a short distance from the carcase, causing the hyenas to stay further away, waiting for him to go. I just keep quiet, watching from my vehicle, absorbing Africa, and making notes. That is the state of play now, as dawn arrives.

Almost at the same time as the sun emerges, shining on the distant Nyaraswiga hill, one single large bird flies near, just above the highest of the flat-topped fever trees. Deep wing-beats, sailing briefly, then wing-beats again: a white-headed vulture. It circles briefly, alights on a low branch facing the kill, and waits, looking around. Nothing else stirs. A few minutes later the bird decides that all is safe enough, and lands next to the zebra remains. The bloated lion just lifts his big mane, then takes to sleep as the better option.

Quietly the vulture starts eating. Even without the rich tapestry of the scenery, it is a wonderful spectacle. Of all the vulture species, only the white-headed can be described as 'majestic' when on the ground: it holds itself in an upright stance, black with a strikingly dignified white head and feathered legs,

White-headed vulture arriving at a hyena kill. (Ineke Kruuk)

with a broad red-and-blue bill and lacking the typical long vulture neck. The white-headed is a loner, and rarely gets involved in a 'vulture scrum' like the other species. It grabs a piece of skin in its broad bill, and starts to twist it off the leg of the zebra, powerfully using its pink claws. I am completely absorbed, noting every detail of the intense behaviour by the strikingly beautiful bird.

While watching, many questions whirl through my head. What is different here, why is this bird on its own, why no others? Is its behaviour different from that of other vultures? Does its different shape have something to do with it? I sit and watch.

Half an hour passes. The vulture eats, the world begins to warm up, more birds stir. Suddenly there is the whooshing sound of air through feathers, and a second vulture arrives, its huge wings spread wide, long neck stretched out, its landing gear extended. Coming to rest a few metres from the carcase, its posture betrays slight apprehension. I am still watching it closely when another breezes into the scene. Within a few minutes many more arrive, from all directions. It is as if someone is signalling and calling in the mob, but I think they all react to those first ones coming down. As soon as several are on the scene they run and hop towards the carcase, and start pulling at whatever meat is available. The lion walks off sluggishly, and the hyenas seem to have lost interest altogether.

The newly arrived crowd are different from the beautiful white-headed vulture that I had been watching until then. The later ones are the 'classical' vultures, very large brown birds, with long, bare necks and heads, black beady eyes, huge beaks and claws. Their numbers soon increase to almost 40, in a heaving, screeching mass. The single, white-headed vulture steps well back.

Once I start counting the birds I begin to realize that what appears to be a heap of just vultures in fact contains two different kinds, apart from the initial white-headed. The bulk of them (30 or more) are white-backed vultures, mostly dirty brown in colour, with a large, tubular black bill. Between them are four slightly larger ones, of the same typical vulture shape but with a more mottled wing colour, a brown back and a tubular yellow bill. They are Rüppell's griffon vultures. In glaring contrast with the earlier scene of serenity, this is a horde competing, two species fighting among themselves and between each other.

It seems that these two screeching and clawing species do better than the white-headed vulture in their very different approach. The latter remains in the background, getting nothing, leaving victory to the violent. Driving home for a late breakfast, I remember that only three of the six regular kinds of Serengeti vulture had been present, but the picture had been complicated enough: they were all targeting the same source of food. Being an ecologist, I wonder – different species should have different resources, they should eat different things. Later, I often see such scenes. I also begin to make some headway with my hyenas, with the descending vultures often helping to find the carnivores. Vultures are among the most prominent competitors of hyenas, and I just cannot avoid them (even if I would want to).

Ever since Darwin's *Origin of Species*, people have been interested in what keeps species apart. It has become a truism that closely related species will only persist together if they behave differently, if they eat different foods or occupy different habitats. Darwin's finches in the Galapagos vary in the size and shapes of their beaks, hence they use different foods. What, then, about these heaps of several species of vultures, competing in the most glaringly obvious fashion possible? Especially since the problem turns out to be more complicated than the one kill I describe here: six species of vulture regularly fight each other. This apart from the other scavenging birds, such as the rare lammergeyer, the

Three species competing: white-backed,
Rüppell's griffon and lappet-faced vultures.

much more common marabou, white-naped raven, tawny eagle and black kite. And, of course, there are the scavenging carnivorous mammals. There are times when all of them are eating from the same carcase.

Part of the answer to my question about competition between vulture species lies in the fact that the carcase of a large wildebeest or zebra is not just a simple, single resource: it is a complicated collection of different foods. All the exploiters behave in a variety of ways, so that they do not always compete as much as appears initially.

One dead wildebeest on the open grassland plains. Hyenas and lions have left after eating much of the soft parts, while some black-backed and golden jackals are still darting around the carcase. There is the usual mass of white-backed vultures, rather fewer Rüppell's griffons, and a couple of white-headed vultures. Commonly there are also several of the huge lappet-faced vultures, a dozen hooded vultures and one Egyptian vulture, all eating their fill.

The white-headed vulture is usually the first to arrive; it is the best-looking one, and it stays well aside of the mass. Later comes the largest of all, the huge, ugly lappet-faced vulture, deep black, its head and upper neck totally bare, pink with fleshy flaps of skin, with a vicious bill that is large even for the gigantic size of the bird. These two both are tearers and twisters, the lappet-faced usually feeding right at the centre of the action, on the carcase itself, very aggressive and getting the best of any spats. Rarely, I have also seen both of these vultures kill small prey themselves, such as a hare or a gazelle fawn.

But the 'typical' vultures in the Serengeti, the birds one sees in caricatures, are the white-backed and the Rüppell's griffon. They go about their business rather differently: they pull out the soft parts, stretching their hideous long and bare necks into openings to feast on the fleshy insides. The bills of all of these four species are vast compared with the bills of the last two kinds, the hooded and the Egyptian vultures. These two are almost similar in size but considerably smaller than the previous four species; each is about one-third of the size of a lappet-faced. Their bills are long and slender, the hooded vulture is dark brown, the Egyptian one dirty white. These two often come to the main carcase when all the others have left. The hooded vultures are like domestic hens, pecking little bits from the carcase itself and around it, the Egyptian vulture taking tiny pieces from the bones. The hooded vulture is the one that I see around the African towns and villages, and they often hang around our house in Seronera. But by far the most common vultures anywhere in the Serengeti are the larger white-backed ones. They constitute the big, heaving crowd, the vulture gathering of nightmares.

I begin to understand something of what enables these birds to exist together, how they share. Somehow, they manage in this ghastly scramble not just the survival of the fittest, but survival for all of them, as part of the enormous Serengeti ecosystem. For me, looking at these birds helps me not just to find spotted hyenas on the endless Serengeti plains, but they also provide insights into how related animals in an ecosystem compete, or survive together. Vultures have problems similar to the striped and spotted hyenas, but solve them differently. It is all part of the same, big struggle.

CHAPTER 15

Flying next to vultures

Rüppell's griffon near its cliffs.

AS ONE DELIGHTFUL PART of my duties in the Serengeti Research Institute I learn to fly small aeroplanes, and like most bush pilots I develop a passion for flying. I love soaring above the plains, seeing the wildebeest herds from up high, watching wildlife from the skies, and landing on distant wild spots in the savannah. This is also what vultures do daily, and often I can admire their effortless performance, their air speeds and stalling speeds, which all add to my fascination with them. Through my practical involvement I become engrossed not just with an aeroplane, but with the flying of these huge birds, with wingspans not much less than half of that of the Piper Super Cub of the Institute.

One day before I started piloting myself, I clamber over large boulders at the entrance of a deep gorge in the Gol Mountains, in hot, dry and desolate country in the eastern Serengeti, far outside the national park. Every so often I look up at the cliffs towering high above me. It is far from any habitation; these are wild mountains that nobody visits. I am here for those cliffs, inaccessible and threatening. They are the sites for a huge colony of vultures, Rüppell's griffons. Just the one species, and it is the one and only cliff colony in the Serengeti.

Coming from the open grassland plains where I had watched prey remains left by the hyenas, I am still puzzled about that problem of the white-backed and Rüppell's griffon vultures. It is those two real 'caricature' vultures, astoundingly alike, that compete hammer and tongs at every carcase. The Rüppell's griffon is somewhat larger and a bit more spotty, but how the two can coexist as two different species, being so similar in habits and appearance, is a serious ecological question. Perhaps the conundrum has a solution in some other aspect of their lives, perhaps in their nesting habits. Hence my presence here, under the gigantic Gol mountain cliffs quite far from the central Serengeti plains and the Ngorongoro Crater where I usually spend my time watching wildlife. High above me are hundreds of vultures drifting in the sky overhead, or sitting on ledges. All of them here are of just the one kind, the Rüppell's griffon.

Repeatedly eyeing the Gol mountain rock faces my heart sinks. I cannot see how I could possibly get any information on what is happening in the birds' nests on the inaccessible ledges, how many eggs or chicks there are if any, and what kind of behaviour is going on in them. The slopes and huge cliffs are dangerous and inaccessible, and the place is hot and hugely inhospitable.

Walking across a lower part of the grass slope underneath the cliffs, many kilometres from anywhere, I notice evidence of civilization. A twisted piece of aluminium, then another. I suddenly realize that I am walking across the site of the tragedy I was told about earlier, one that was very relevant to my presence in the Serengeti, and which took place only a few years ago, right here at these bird cliffs. What happened was this.

A father-and-son team from the zoo in Frankfurt, Bernhard and Michael Grzimek, were here to make a unique contribution to conservation, and to the understanding of the migrations of the herbivores. They surveyed the Serengeti animals in the 1950s in a small German aircraft, a Dornier. They were the first ever to do such a survey, in an attempt to find where the migrations of all the wildebeest and zebra and others went, year round. The aim was to establish proper boundaries for a national park. Early in January 1959, Michael was piloting his plane here near the Gol Mountains, near the bird colony right where I am standing now, when he hit a griffon vulture. His right wing was damaged and the plane nose-dived, crashed and disintegrated, close to the cliffs. He was killed instantly.

Michael, 25 years old, was buried on the rim of the Ngorongoro Crater; his grave overlooks the crater and is now visited by many. His father Bernhard returned to Germany, where he wrote a runaway success book, *Serengeti Shall Not Die*, and produced a film. From the resulting enormous public interest and funds the Michael Grzimek Memorial Laboratory was built, and with that, the Serengeti Research Project established.

With melancholy thoughts about what happened here years ago, and about the possibilities and mostly impossibilities of watching vultures in this place, I clamber back into my Land Rover and return home to Seronera. From what I have seen, I believe that the Rüppell's griffon colony can only be counted and

Rüppell's griffon colony on the Gol cliffs, Serengeti.

observed from an aeroplane. I also think of those pieces of aluminium below the cliffs. Aerial observation should be feasible, but impossibly dangerous. I go back to my hyenas, but keep thinking about vultures.

Now, more than two years after the small Michael Grzimek laboratory was built, the Serengeti project has taken off, with keen interest from all over the world. National Parks director John Owen has pushed and cajoled prominent scientists, universities and institutes, and persuaded industry to open its cheque-books. It now is a Serengeti Research Institute, there are more than a dozen scientists and many assistants; facilities have blossomed, and we have three aeroplanes.

One of our activities involves expanding the Grzimeks' work, to establish exactly where and when the huge migrations go inside and outside the national park, to establish numbers of animals and why there are so many, and to measure details of climate and vegetation in all parts of this huge area. That is why I learn to fly, to check rain gauges in remote parts, to count animals from the air, also of course to find hyenas, and to help other scientists with their work.

When piloting one is very much aware of the vultures, especially when above the large herds of wildebeest. The birds are a major hazard to any pilot, who has to be constantly on the lookout. I have several close encounters. But at the same time, flying provides an opportunity to really look vultures in the eye when they are high up, drifting way above the plains. To be up there is a magical experience, alone in the small aeroplane when motoring along, engine throttled right down and flaps down, as slow as I dare. I see a large white-backed vulture in all its majesty sailing quite close, with its big black eyes that always manage to look furious, its huge brown, fingered wing stretched towards me. It seems as curious about me in my plane as I am about the bird.

My main interest may be the predators, the hyenas and lions, wild dogs, cheetah and leopard, but the vultures never quite let go, and inevitably it gets me back to the old problem of the similarity in behaviour and feeding of the two main kinds of vulture. Perhaps their differences in nesting behaviour when they are away from the kill is relevant. The larger, more dominant Rüppell's griffon only breeds on the inaccessible ledges of high, steep mountain cliffs in the formidable Gol Mountains, in one large, quite dense colony. White-backed vultures nest on top of the ubiquitous, flat thorny acacia trees, much more dispersed but also inaccessible, anywhere throughout in the central Serengeti. So I need data on the birds' numbers, their nesting habits. I would like to know how many nests there are and where, when the birds breed and how they manage, and what happens to the chicks. I am still dreaming that with access to an aeroplane I can have a go at it.

The Grzimek tragedy is never far from my mind when flying the small Piper Super Cub in that area of the Gol cliffs, mountains that belong to the vultures. The cliffs are over 300 metres high, and the huge birds are everywhere. In theory one can, indeed, fly very slowly in this light plane with a large engine, cruising comfortably even at only 40 knots. But to be able to count the nests and look

Rüppell's griffon and white-backed vultures, drying.

into them for eggs and chicks one needs to be close to the cliffs, with the danger of collisions and down-draughts, with little space for avoiding oncoming avian traffic. I am very aware of my limited flying experience and, as it turns out, I never have the necessary confidence to get close enough to those cliffs and deep ravines while piloting myself, simultaneously counting and looking at nests.

Fortunately, several years after my hesitant beginnings of flying near vultures, my colleague Hugh Lamprey, then director of the research institute and a more experienced bush pilot, offers to help. He will pilot the plane, and I will do the nest observations. We try it hard, with vultures whizzing past the plane at close quarters, in and out of narrow ravines with ghastly down-draughts: it is hair-raising. Soon even Hugh decides that retreat is the better part of valour. We land on a small, high plateau of the Gol Mountains and, feeling shattered, we spend the rest of the morning in a splendid pool under a waterfall.

The next brave man to try to sort out my vulture problems is Colin Pennycuick, a research colleague who is a natural pilot with many years' experience after a flying career with the RAF. Flying with him the vulture attempt almost ends in disaster. He pilots, I am the observer and we count many nests, but even that is only a small proportion of the total colony. It is insufficient and before long he, too, thinks this kind of flying is too dangerous.

After giving up on the vulture cliffs and judging that further attempts would be lethal, Colin and I decide to fly on, and just out of curiosity have a good look at the shores of nearby Lake Natron, a huge lake and totally forbidding. Its shores are muddy salt flats. Nobody ever comes here – it is too far from any road or habitation, it is incredibly hot, an eerie, fabulous but smelly landscape, dominated by white salt and hundreds of thousands of flamingos. The shore of the lake looks magically interesting when we make a low, slow pass. From the plane we can see many dried corpses of birds and mammals that died in the terrible, smelly brine along the shore. The salt flats themselves look smooth and hard, the aeroplane has large, soft tyres, and we decide to have a go at it, and land.

But Colin's expert landing goes wrong: the salt layer that looks so hard and easy from the air is in reality only a thin hard cover over soft mud. As soon as the wheels touch down he realizes that there is trouble, and Colin instantly decides to pull out at full throttle, rather than stop. If we had landed, it could well have been our end. During the unexpected and slow take-off, with the mud dragging the undercarriage, the landing gear collides with a boulder. It bounces up and hits our tailplane. We find ourselves in the air again, above one of the most hostile landscapes in Africa, with one of the plane's wheels damaged and undercarriage bent right back. Looking over my shoulder from the passenger seat I see that the boulder badly buckled one side of the tail (which controls vertical movement of the aeroplane).

Somehow, Colin keeps the plane in the air and expertly under control, by leaning heavily on the joy stick. We are about one long hour flying from Wilson Airport in Nairobi, where pre-warned fire engines await us next to the runway. Colin lands the plane smoothly on one wheel, and only when we come to a near stop does it fall gently onto its wing-tip on the runway.

The Piper Super Cub at Lake Natron.

That is the end of my attempts to approach the Rüppell's griffon colony from the air. The birds do give up some of their secrets in later years, when David Houston comes to do his PhD with me on the Serengeti vultures. He spends many days watching about 150 birds on their nests inside a deep gorge in the Gol Mountains, sitting on a slope opposite the cliff, and he walks and climbs many of the other areas. His is a landmark study, and he estimates that the total colony of Rüppell's griffon vultures has well over 1,000 nests. They have their chicks, one in each nest, exactly at that time of year when the large migratory Serengeti herds of wildebeest and zebra are usually closest to the Gol Mountains. This still means that later in the season many of the birds have to travel more than 150 kilometres from their nests each day to reach their food supply.

David also studies the nests of the white-backed vultures on top of the thorny acacias in the central Serengeti. Scattered everywhere over large areas, their chicks hatch each year about three months later than the Rüppell's griffons, also at the time when there is most food nearby. The tree nests of white-backs are unreachable from the ground.

In this case, fortunately, the use of our aeroplane proved to be the saving grace. As an exciting piece of work, every week I slowly fly the aeroplane just above the treetops with windows and door open, wing-flaps down. David, the observer behind me in the little plane, can see and count the eggs or chicks in the nests, when the concerned parent bird stands up at our approach. With such data, he is able to build a detailed picture of vulture numbers, breeding season, breeding success and other aspects of their life.

The different nesting habits of these vultures explain at least partly how the two species can survive next to each other, despite their horrendous competition over carcasses. The larger, dominant Rüppell's griffon needs the benefit of steep cliffs with strong winds for take-off. It starts its foraging trips by just falling off a ledge. The rock faces are a long way from the fleshpots, and the birds' commuting must cost a lot of energy (despite their fabulous soaring and gliding abilities). On the other hand, the slightly smaller, white-backed and underdog vulture can nest much more closely to food, on one of the many acacias, and therefore it can make savings in travelling energy expenditure. As a result, it is the white-backed vulture that is the most numerous overall.

Taking off from the ground is hard work for a large and heavy bird, and often it needs a considerable run before a vulture becomes airborne. They are vulnerable then, and I use that when catching and colour-ringing a number of them. I want to get resightings of ringed vultures on carcasses, to establish what the birds' range-sizes are. So I catch several white-backed vultures and Rüppell's griffons when they have a very full crop, racing after them in a vehicle when they try to take off from the open grassland; somebody on the running board of the car jumps and grabs the bird when we come alongside. In a more dignified and professional manner, David Houston improves on this exercise by catching the birds with a canon-net at carcase sites. As it turns out, there are so many vultures in the area, thousands of them, that it is rare to ever meet one of the colour-ringed birds again.

Of all the vultures that I see around kills, the two smallest – the hooded and Egyptian vultures – are also the ones that often hang around African villages. At their size they do not have a problem with taking off from ground level, but they are not such superb gliders as the larger ones. Hooded vultures make their nest inside the crown of large trees, and Egyptian vultures are cliff-nesters. And this last one shows an almost sophisticated kind of behaviour that separates it from all others.

The Egyptian vulture is one of few birds anywhere in the world that uses tools. When I read about this, I want to see it for myself, one day when I am out on the grassland plains of the Ngorongoro Crater. The result is a more beautiful sequence of events than I could ever have hoped for.

Some days earlier, I had found a single ostrich egg just lying about. Quite often ostrich hens are caught short while far away from a nest, and they lay and abandon an egg in the middle of nowhere. Early in the morning, I put the huge egg prominently on short grass, not far from the cabin where we are staying in the crater. Sitting quietly at a distance in the Land Rover, I watch what happens. As usual, there is all sorts of coming and going from both birds and mammals.

After about half an hour, a pair of golden jackals ambles past. They sniff at the egg, then one of them starts rolling it with its front legs, going backwards. Perhaps they were going to cache it somewhere, but I'll never know what they would have done with it because they are rudely interrupted by two spotted hyenas, who notice from a distance that something worthwhile is happening.

As soon as the hyenas move in, the jackals leave. For more than 20 minutes the hyenas try to get their strong jaws around the giant egg, pushing each other out of the way. I cannot help laughing when they don't seem to make the slightest impact: the ostrich egg with its remarkably strong, thick and smooth shell appears to be too large, too solid and too impregnable, even to hyena jaws.

Curious, a white-backed vulture lands and watches, sitting quietly a small distance away, waiting for spoils. Finally, the two hyenas give up, walk off and lie down some 100 metres further on. The vulture flies off.

This behaviour of the hyenas is intriguing, as I cannot believe that their solid jaws cannot crack an ostrich egg. In fact, in the Kalahari desert my student Gus Mills shows that brown hyenas in the Kalahari have no problems with an ostrich egg and just bite it open, as their jaws are slightly larger even than those of spotted ones. However, as it turns out spotted hyenas may not be able to open an egg themselves, but 'they have ways'.

As I watch from my vehicle, an Egyptian vulture flies past, circles and lands next to the ostrich egg. It walks around it, looking at it intently. Soon it flies off, but lands again about 20 metres further. The next thing I know, it comes walking back to the egg, carrying a stone that is larger than its own head: later I weigh it as almost half a kilogram. The bird is a splendid sight, lifting the stone to well above eye level, then dropping it on the egg with an audible clang. It repeats the exercise six times before it is rewarded by a crack and a small hole in the shell. Clearly, a tool for the job.

Sadly and undeservedly, just as the Egyptian vulture is dipping into the egg contents, competition strikes. Two hooded vultures mysteriously arrive, seemingly from nowhere, obviously just at the right time. After a short skirmish the two hoodeds are feeding, leaving the Egyptian vulture moping, sitting at a distance. But alas for them, as egg-white drips from the hoodeds' beaks when they look up, they see the same two spotted hyenas walking back again to the scene, clearly attracted by the vultures' activities. That is the end of the vultures' luck. One hyena picks up the egg with its jaw inside the hole, cracks open the shell and, while it is licking the spilled contents, the other carries off the shell with most of the yolk.

Tool-using Egyptian vulture opening an ostrich egg.

I am delighted with this gold mine of incidents of competition, in just one single observation. It makes me realize how much of this kind of happening must be going on unnoticed. Apart from the interesting vulture and hyena behaviour, one can also see it as a tiny episode in the life of an ostrich.

On a later occasion I try to repeat the experiment by putting out another abandoned ostrich egg on our bird-feeding site at our Serengeti home in Seronera. Here we are visited by the occasional Egyptian vulture, and I leave the huge egg conspicuously on the grass nearby, in the hot, midday sun. Small birds feed around it, while I keep an eye on it when working on my typewriter on the veranda. This time, unbeknownst to me, the ostrich egg must be many weeks old.

After half an hour in the hot, tropical sunshine, the egg bursts with a huge explosion. Pieces of its thick shell hit me on the veranda and against the windows, leaving a ghastly, rotting egg smell over a large area. I reflect that if an Egyptian vulture had come and struck the ostrich egg with its stone tool, it could easily have been killed. Obviously, there is quite a danger even in a stray ostrich egg.

Such hazards, however, are small compared with the strains that competition in general puts on the existence of these birds, in fact on all the vultures. Apart from the vicious scramble with other vultures and birds, they also have to contend with mammalian predators, with hyenas and lions, jackals and many others. It is an unambiguous, nasty combat for survival, not just some indirect competition for food, but a genuine fight. Perhaps it is a struggle such as that of the vultures that has given Darwinism, with its 'survival of the fittest', a bad name.

Unfortunately, it is not just other wild animals that bother vultures. More serious threats come from quarters where I had not expected them at all, from ourselves. Many years after I watched the rich diversity of vultures in the Serengeti I find myself on a quite unrelated activity in Europe, photographing orchids underneath a mountain range on the island of Crete. There, intent on the beautiful flowers in front of me, my heart leaps when I suddenly hear that familiar whooshing noise from the air again.

There it is, a vulture, this time a Eurasian griffon just above me, which is very similar to the familiar Rüppell's griffon in Africa. It has its landing gear out, heading for a place just over a small hill in front of me. I forget about the orchids, battle my way through the maquis to try to follow the vulture, panting, expecting to find a kill, as I would have in Africa. This time, when I get to an open grassy site on the other side of the hill, I am confronted by the harrowing sight of a sick bird, sitting miserably, taking little notice of my presence. I feel sick myself, dazed with disappointment and furious, when an hour later the vulture dies, right there in front of me – a victim of poison. I write letters, but the authorities shrug their shoulders.

Vultures must be just about the easiest of all birds to kill by poisoning, and massive numbers have succumbed this way in almost every country where they occur. It beggars belief, because it is so totally pointless. Other disasters that befall them at least are accidental – that is still bad enough, but understandable. High-voltage pylons are one such major and man-made cause of

Griffon vulture in Crete, poisoned, and moments before it dies.

death, especially in South Africa, with birds spectacularly electrocuted in a ball of flames. Fortunately, recent changes in the design of pylons are putting a stop to this.

The most serious threat to these majestic birds arises in the 1990s, starting in India but spreading far and wide, and fast. Poisoning, accidental or otherwise, is caused by farmers using a drug called diclofenac for their cattle (also often used medically by ourselves). It causes kidney failure in vultures, even in very small doses. It is now banned for use in cattle in many countries including India, but not before three Asiatic species of vulture have been virtually wiped out. Dangers to the African vultures are just around the corner.

Despite their huge size and all the violence that the vultures create among themselves, despite their neat apportioning of food resources between species, I am only too aware that these magnificent creatures are hopelessly vulnerable. I worry. In many people's eyes they are the lowest of the low, but the sight of these huge birds, both on the ground or in the air, has a breath-taking, bewitching grip over me. Of course I am terrified of hitting them when I am up there, in my small aeroplane, and I wish I could pilot my plane as well as they fly. Nevertheless I can see them as a magnificent presence, a very visible and impressive part of the wildlife community.

CHAPTER 16

Chasing dogs on Darwin's islands

Feral dog on Galapagos lava.

L IVING IN THE Serengeti and Ngorongoro with the hyenas is a privilege like no other, and I can only think about it in superlatives. I think of my incredible fortune having the animals, the environment, the space, the people, the science. Yet there is one other area in the world that often plays through my mind, which is endowed with fame and natural treasure in a way that invites comparison with Africa and the Serengeti. The Galapagos. I wish I could experience it, contrast it with what I already have here in Tanzania, visit it as the cradle of evolutionary theory. Almost unbelievably, I am going there, in a move that arose from a totally unexpected corner.

Around our Seronera house in the Serengeti in the rocky outcrops, the *kopjes*, and throughout the plains on rocky hills, we have lots of the curious animals called 'hyrax', or 'dassie' in South Africa. There are two very similar species, often living together among the rocks, and there is one other living in the bulky fig trees. They are somewhat larger than guinea pigs, tailless and closely related to elephants, jumping about in the trees and over the stones on their rubbery feet, lovely and noisy. They get into houses and eat the plants, and they may sleep in the engine of parked cars. In Swahili they are called *pimbi*, and a good friend of us, also living here in Seronera, is making a PhD study of them. Henrick Hoeck is hence called *bwana pimbi*, but to us he is just 'Pimbi'.

Pimbi originates from Colombia, and his heart is in South America. When he finishes with his PhD on hyraxes in the Serengeti, the lucky man is offered the job of director of the famous Galapagos Research Station on Santa Cruz, one of the Galapagos islands. As one of his large challenges there, he decides to tackle the major problem of non-native, introduced animals, which are causing huge damage to the native fauna of tortoises, iguanas and all the birds and other animals unique to those islands. It is this fauna, now threatened, that inspired Charles Darwin to come up with the answer of evolution.

Over a beer, Pimbi convinces me that my experience with the role of hyenas and predation in the Serengeti is useful to him for understanding the effects that introduced animals, especially domestic dogs, have on native Galapagos species. They prey on many, including the huge lizards, the iguanas. He needs to get ideas about how to deal with these incomers. For me it is a chance not only to experience the Galapagos, but also to see other pack hunters different from hyenas in action, other adaptations of a carnivore in a new environment.

And here I am on the islands, and have been now for a few weeks. My present stage is a small, low hill, with a couple of rocks on the top, in a moonlike landscape of recent lava. It must be one of the most inhospitable, blackest landscapes in the world, the southern shore of the largest Galapagos island, Isabela, almost on the Equator in the Pacific Ocean. The sun is blasting me while I seek protection by backing into shade against the rocks. Protection, this time not from the intense heat, but from four, almost entirely white, large dogs bent on attack. They are feral dogs as wild as they come, ownerless and extremely aggressive, defending their patch against me, the intruder. They bark, snarl and growl only a few metres away, and I am facing them with my back against miserably small rocks. I am rather worried.

I have been warned about these wild animals by previous visitors to the island, but against advice I declined to have a firearm with me. I have to sort out exactly what these dogs are doing here, and my feeling was that being armed would set me off on entirely the wrong footing. I have little time to regret that decision, as things are happening too fast. Tails up, the pack is coming for me, led by a large male which earlier I had dubbed 'Ten-past-eight', from the way in which he holds his ears. In my shorts I feel vulnerable, I am on my own, very slow compared with my attackers, and there is no cover of trees or climbable rocks for escape within miles. Facing the dogs and without thinking, I more or less instinctively pick up a sizeable lava stone and throw it at the front dog, hard and fortunately quite accurately.

It yelps, and all four of them turn tail and run, stopping and barking at me from a much more agreeable distance. A couple more projectiles firmly decide the skirmish in my favour.

That was nasty. Yet I cannot help but rather like and admire these rogues. Mostly white, the dogs are an audacious, noisy and very conspicuous intrusion in the black lava fields along this unfriendly coast. There are many of these feral animals in what is now a totally uninhabited area, descendants of dogs from an abandoned, early-twentieth-century convict settlement, and a canine anathema to the native animals. It is not known, though, how much of an impact these intruders really have, and on which of the native species. I am trying to find out.

The situation is immediately obvious when I land here a few days earlier, deposited by my colleagues from the research station. After an overnight trip on a small fishing vessel from the research station on Santa Cruz, our landing party heads for the difficult lava coast of Isabela. I am dropped with an assistant and left in this area, which is known to the local fishermen for its vicious dogs.

Caletta Webb is on the south-west point along the shore of Isabela. There are no roads or tracks, and it is many miles from anywhere. There are some remains of an American base from World War II in the Pacific, otherwise nothing that is reminiscent of civilization. It is difficult to walk about, because of the fiendishly sharp, fresh lava everywhere, large flows of lava rocks with the appearance of black bread but with the face of millions of needles. The distant volcano Cerro Azul dominates the black desert scenery, and if ever a little rain would come down, it would disappear straight into the ultra-porous sand and rock. Apart from a few cacti there is virtually no vegetation, no shade. Around is the wide, wide ocean, and it is hot, very hot.

Arriving from the sea, I am very aware of the fabulous richness of these waters. On the little fishing boat we are accompanied by whales and dolphins, by petrels, frigate birds, boobies and other birds fishing around the ship. But life on the island is something else. Getting onto it has to be attempted with a Zodiac dinghy on those ghastly rocks and with an impressive swell on the sea. It needs masterly boatmanship from one of the scientists, Howard Snell, who is an iguana specialist. He is showing off his skill, with a fast approach towards the rocks on the crest of a wave, then a jump by one passenger (me) when the dinghy almost touches the rocks, the little boat withdrawing again on that same wave. The next approach sees a parcel being thrown across, then a backpack, a tent, a container with water, and finally a jump by my assistant Fausto: one at a time for each approach. Mercifully, there are no mistakes.

Looking around, I see that I am an actor in some marine drama, with an abundant audience close and all around. On the cliffs above the waves, often on the very tops and facing the ocean, are many hundreds of huge, dark grey, round-faced and crested reptiles, of all sizes but some of them well over a metre long. Marine iguanas are the most conspicuous sign of life here, and although they are facing us, the animals seem to take hardly any notice of our activities. Neither do we of them, as we are occupied with our own difficulties of landing here. The iguanas, silent dragons, just sit, sunning themselves alone or in dense crowds. Occasionally some swim out, dive and disappear for ages, feeding on algae on the bottom.

Then, during all the excitement of the landing and having spent only minutes on Isabela soil, we spot the first dogs. I expect them to be aggressive. Some 50 metres away they stand in a pack of about a dozen, large and white in a black landscape, with their tails up, and barking at us. They have an eerie

Marine iguanas on Isabela, with two blue-footed boobies.

beauty, and they look attractive because of their domestic familiarity to me, in this utterly hostile environment. How could these animals possibly cope in this hot desert? There is virtually no edible vegetation, and the fauna consists of just marine iguanas, some fur seals, the odd Galapagos penguin, a cormorant and some smaller birds and lizards.

Standing next to our meagre possessions on the utterly bleak shore, I notice three of the dogs walking off towards the cliff edge. Tails curled up in an effortless trot, once they are some 30-odd metres from the others they accelerate, and before I realize what is going on they are tussling over a medium-sized iguana, tearing it between them. Other dogs immediately join in. I watch them with their probably daily fare, four of them tucking into the hapless victim, soon each dog having its own piece.

After Fausto and I have set up camp in a small, sandy depression, I sit down to make some notes, and think about the project. Funded by the Frankfurt Zoological Society, my objective is to find out what the effect of the feral dogs is on the native fauna, and to advise the Galapagos conservation management team whether to let them be, or otherwise. I now realize what a challenge this is, in this immensely hostile landscape. Virtually nothing is known about the ecology of these animals, and where do I start, in my short period of a few months available? In fact, it later becomes even more of a challenge. Once I begin finding out about the dogs, about their incredibly interesting lives on the lava, once my sentiments about the animals run against rational thoughts about the need for management, it is difficult to be the objective scientist.

That very night, the first after my arrival, the dogs do not try to ingratiate themselves with me, although I have come to sit in judgement upon them. It is hot, dark, and while trying to get some sleep in my small tent I hear them, barking and howling nearby. Choruses of barks move across the night, in a way that takes me back to African nights with hyenas, when large packs chase other large packs in group territorial clashes. Even when I am still thinking about this, there is a jerk on one of the guy-ropes of the tent, and peering out I notice that my bathing trunks, which had been hung out to dry, have gone. In the morning I find them nearby, torn to shreds. The dogs made their point, they rule this area where, in daytime, there is no shelter, no green, no easy way to walk about, where the sun blasts in supernormal temperatures, where there is no respite from the relentless environment.

How exactly these dogs got here is uncertain, but they did arrive on the coat-tails of mankind. Wherever people settle, they bring their pets. On even the remotest islands, anywhere in the world, lighthouse-keepers, sailors or colonists decided that they cannot do without their companion cats and dogs. Unfortunately in the Galapagos, once they arrived, many of those companions decided that they could do perfectly well without their lords and masters. They found rich pickings among the fauna on those islands because, having evolved in the absence of predators, the native island inhabitants have little defence or fear. For a cat or a dog, an island somewhere in the Pacific or Atlantic is crowded with the proverbial sitting ducks. The friendly companions turn into exterminators.

Galapagos dogs are no exception, and local people say that the dogs' ancestors were actually put there for the purpose of extermination. In the late seventeenth century, British pirates in the Pacific made life miserable for sailors, and to maintain their nefarious activities far away from home with fresh supplies, pirates stocked various islands with goats for meat – as they did here. But in 1685 the Viceroy of Peru, in a crafty move to make the pirates' life uncomfortable, released dogs on some of these islands to kill off the goats.

Whether it is the offspring from these animals that survived is not known. More recently in the nineteenth and early twentieth centuries, colonies of Ecuadorian convicts were established here, on islands where fleeing was not

Marking the group territory.

an option. These convict settlements had dogs, and when provisioning of these places eventually became too difficult, the wardens left with their convicts, without bothering to take the dogs. Several seafaring travellers in the nineteenth century, visiting the Galapagos, commented on the presence of 'owner-less' dogs.

Whatever their origin, the animals I see here obviously had been in the Galapagos for quite some time and many generations, and interestingly, they had changed. Quite appropriately, Darwinian natural selection had taken its toll. The feral dogs are significantly different from the average domestic or pie dog, which I discovered by walking the village streets and visiting *fincas* (farms) on the islands, where I measured and estimated domestic dog sizes and colours. The dogs that I find feral are some 10% taller than the average domestic dog, almost as tall as the Serengeti hyenas. Dogs with a predominantly white coat I find rare among domestics, but the feral dogs are almost exclusively white. It must have come through harsh natural selection, which to me is hardly surprising, after being out on the almost inaccessible, waterless black lava fields, in the burning sun.

However, despite their long presence here the dogs have not come to a harmonious coexistence in a predator–prey balance with the local fauna, unlike what I find with hyenas on the Serengeti plains and Ngorongoro Crater. As other scientists and local people tell me, dogs in the Galapagos settle in an area in numbers, and for unknown reasons, after some years they move on again, starting up somewhere else. I think that those 'unknown reasons' are very likely the dogs' running out of food, that is the extinction of their prey such as iguanas. The dogs killed their own golden goose.

The question now is, how much long-term harm will they do to the precious populations of indigenous animals? Can the Galapagos sustain such predation, or will the island fauna, that provided so much material to Darwin's theory, be irreparably damaged?

Such thoughts become even more worrying every time I am reminded of the unique, fabulous, almost primeval environment here. One evening on Isabela, sitting in front of the tent in the dark, I notice a bright glow from over the shoulder of the large volcano Cerro Azul, which towers over the entire area. The next day and night Fausto and I clamber and walk over there, to be confronted with the magnificent eruption of a small volcano. The night sky is aflame, with a noise as though several trains were descending on us. Large molten chunks are thrown high up, and a stream of lava winds its way down. The heat, even at some 300 metres away, is unbearable.

I am awestruck. What perhaps impresses me most is, at night, the numbers of mesmerized insects, locusts, bugs, beetles and others that hurl themselves towards and into the inferno. What waste, what maladaptation! Somehow, it makes me aware that the Galapagos is nature at its most basic and untouched. This is where it all happened, and where all is still happening.

My small pilot study may not provide any immediate answers to the questions about the sustainability of the dogs' presence, but it will be a beginning. I am getting first clues, some hard, scientific data on their impact. What I had not anticipated, though, was that I am also coming to see the dogs as fascinating predators in their own right, adapted to the unrelentingly

Eruption on Isabela.

harsh environment in which they find themselves. Even their physiology is astonishing.

Unwittingly, we had pitched the tent in a place that was particularly important to the animals. It does not look any different from the rest of the area, consisting of harsh lava with sandy patches, accessible only with great care, and there is no vegetation. But it happens to be just on the boundary between the territories of two packs of dogs, and they let me know it. Several times, in the middle of the night or in the early morning, big battles are going on, with groups of several dogs chasing others, barking incessantly, yelping when they make physical contact, occasionally indulging in long howling sessions. These are group-living, group-hunting social predators. It is as if I am back among my clans of hyenas.

One morning on Isabela I walk and scramble about a kilometre from the tent, arriving close to the den of one of the groups of dogs, in a cave under a ledge in the lava. A suspicious couple of dogs move away, leaving their headquarters and whatever is inside it to me. In the boiling hot, glaring sun I peer around, collecting dogs' faeces for later analysis, finding two dog carcases while in the distance one present resident barks at me. But the dominating feature around the den is not dogs. It is the multitude of carcases of marine iguanas, scores of them scattered everywhere. It is a knacker's yard. I count, and I measure.

Walking back, still impressed and depressed by all this, I spot a pack of four dogs, one of them my friend Ten-past-eight. They haven't seen me, and I crouch to keep out of sight. In a determined gait they head straight for the coastline, to a small black beach between the lava cliffs. And there just in front of me, they solve a question I had been asking myself. All four of them walk out into the waves of the Pacific Ocean, and they drink, lapping up the seawater. This was staggering, to see a mammal just like I am myself, drinking salt water from the sea. Heaven knows what its kidneys must be doing.

Also the social organization of the dogs fascinates. They move about in packs, not just all over the place, but packs that stay within prescribed limits. One of these limits is close to the tent, and with ferocious exchanges there the dogs defend a group territory against neighbours. 'My' dog group contains some 20 animals (excluding a few small pups inside the den) who go around in groups, or sometimes on their own. Which again is remarkably similar to a clan of spotted hyenas.

The more I see of the dogs, the more intriguing they become, clearly possible subjects for a quite unique scientific study. But agonizingly, I am also confronted with their impact on the other animals here. It is the dogs' effect on those which evolved here earlier in these fabulous Galapagos islands, on the marine iguanas, on the fur seals and the birds. The mortuary of iguana skeletons is like a bad dream, all those remains of animals of a species that is unique in the world. Perhaps the stark environment magnifies the dogs' deeds, but the consequences of their presence are glaringly obvious.

Marine iguanas clearly take the brunt of the onslaught. They cannot measure up against the dogs, they just sit there, being slaughtered. As an innocent island-species, they have no adaptations to protect themselves against predators. The prehistoric-looking reptiles sun themselves, motionless on the rocks, recovering some heat after having dived into the cold ocean waters to feed. Each individual looks as if it has survived for ages.

It does not come as a surprise that the marine iguanas are very slow at reproduction. Iguana specialist Howard Snell tells me that females have to be several years old before they start laying, and then they produce only two or three eggs per year. Which, incidentally, they bury in sand, to the delight of the dogs, who quite easily find the eggs. Egg predation is another way in which iguanas are exposed to the feral nuisance, on top of the more direct killing of the animals themselves.

One thing that strikes me when walking near the dogs' den between the abundant skeletons of marine iguanas, is that so many of these remains are of very large individuals. When I measure the skulls, it is clear that the majority of them are of the very largest iguanas, of more than a metre long. Compared with the sizes of iguanas that are sunning themselves on the rocks, of which many are small, this shows the large ones of a metre or more to be at much greater risk.

Are the beautiful, gnarled, largest iguanas, the old males, especially targeted by the dogs, or are they particularly vulnerable in some other way? The best thing to do for answering a question like that is just to watch what happens, to see the reactions of the reptiles to the dogs – or to myself.

Through my binoculars I see Ten-past-eight, the easiest dog to recognize, walk along the coastal cliff in the early-morning sun, amazingly close to the sunbathing reptiles. Some of them hardly get out of his way. It is especially the largest and oldest ones that are tardy to the point of recklessness. The dog, or as I find later, I myself, can almost touch the large iguanas before they decide to shift (and then they go very fast). The smaller ones and especially the baby iguanas move when the dangerous dog is four metres or more away. But the very largest, conspicuous males hardly bother to run, they stand their ground – badly adapted to the threat of predators.

Clambering the sharp rocks late at night, with my electric torch in my hand and surrounded by the noise of the surf, the only marine iguanas along the coast that I can find are the few impressively large males sitting on prominent rocky outcrops. They are the ones left on the rocks at night. They are still there proclaiming their territory in the dark, while all the females and youngsters hide in crevices. Darkness is the time of most dog activity. The behaviour of these iguanas neatly explains why it is that the large iguanas are the ones that get caught.

There are not many of these largest reptiles left along this coast. In daytime I still find numbers of smaller-sized ones, but I worry that, with the very prominent presence of the dogs, the future of these reptiles here on southern Isabela may be far from rosy.

At night in my tent, after a few days of watching and counting, I make some very rough back-of-an-envelope estimates of what is going on between the dogs and the hapless population of marine iguanas. My inspired guess is that dogs take about 27% of the iguana population each year, and that does not take into account all the eggs that are dug up by the predators. There is no doubt that, with the very slow reproduction of the iguanas, a population of these reptiles cannot sustain such losses.

Not surprisingly, I hear that other visitors reported from several coasts on other Galapagos islands where there were dogs not many years ago. Marine iguanas also used to be abundant in these sites. Now, both iguanas and dogs have totally disappeared from these places.

The canine misdeeds in Galapagos don't stop at the marine iguanas. Another species, the closely related land iguana, has suffered much worse. Land iguanas are somewhat different from the marine ones, about the same size but greyish, green and yellow, instead of the dark, almost black and reddish marine iguanas. The land iguanas are and were far less common than the marine ones, numbering a few hundred in the most recent estimates, compared with the many thousands of marine iguanas. They are also more vulnerable as

Old male marine iguana.

Marine iguana portraits.

individuals, because they have no easy crevices and steep cliffs to hide in. Land iguanas dig shallow holes in the sand as dens, which provide little protection against ferocious dogs. Scientists studying land iguanas have seen entire island populations exterminated by dogs in a very short time, such as on the central Galapagos island of Santa Cruz.

The list of victims of feral dogs also covers other precious species. The famed Galapagos tortoises, of which only a few are left, suffer by losing their eggs and very young ones to dogs (as well as to feral pigs). The unique, endemic Galapagos penguin has the unfortunate habit of roosting along shores in vulnerable places, and I find quite a few of their remains in several of the dogs' scats. Fortunately, most of the penguins nest on the smaller islands where there are no dogs – and the same is true for the unique Galapagos fur seals, which are a frequent prey for dogs on Isabela and elsewhere.

In calculating the dogs' balance sheet, I have to admit that they also have a positive side. They kill other, misbehaving and imported exotics, of which there are many such as cats and rats. In my blinkered approach to dog problems so far, I have hardly even mentioned all the other introduced animals that are causing headaches to those of us who love the Galapagos.

It is astonishing how many of mankind's companions have been transported and released here, an almost endless list of plants, insects, birds and mammals, mostly domestic. In the dense bush of the hills of the island of Floreana I hear feral domestic cocks crowing everywhere, far away from human habitation. Entire hillsides of several islands are covered in guava trees (which do not belong here), and on the larger islands there are totally wild goats, pigs, cattle, donkeys, horses, cats and rats. Many of them have a much more profound impact on the vulnerable Galapagos environment than the dogs do.

Some of these nuisances feature prominently on the dogs' list of prey. I find cat remains in their scats, and other scientists have reported that cats also do enormous damage to marine iguana numbers, by killing all the young ones. Where there are dogs, one rarely sees evidence of cats, so dogs may well keep feline numbers low.

Similarly, there is the ubiquitous goat, which has devastated some of the islands. They are not a problem where there are feral dogs, because goats are an easy and highly preferred prey of dogs. Wild cattle are common in the highlands of Isabela, quite far from where I studied the dogs and iguanas. I hear from the local people, the Galapageños, that dogs hunt cattle in large packs (like hyenas hunting zebra). I never saw that myself, but there is cattle hair in the dogs' scats that I collect.

Yet despite the potential benefits that dogs may bring to the beleaguered islands, and the intrinsic interest of the dogs themselves, these are hardly enough to balance their ecological crimes. Even when I keep in mind the scene of those dogs walking into the sea to drink and their fascinating clan battles, when I see the battlefield of iguana remains in the area close to my tent on Isabela, I have no doubt that these introduced ferals just have to go, and I advise accordingly.

Removing dogs, however, is not easy. Shooting or catching is difficult in this often inaccessible terrain; they learn quickly and many will escape. It appears that poisoning is the only efficient alternative, a method successfully used on islands near New Zealand. Without taking this any further but as a pilot experiment to see if the dogs could be attracted to bait, I put out some fish buried in sand. This does not expose any of the endemic Galapagos animals. The Galapagos hawk, a species of buzzard, is around here, and it eats carrion – one would not want that bird to eat a poison left out for dogs. Even land iguanas might eat carrion and endanger themselves in a dog eradication campaign.

The dogs readily find the bait and eat it, and also when I leave a plastic basin of fresh water on the rocks it is immediately emptied by them (they obviously prefer this to sea water). With such options I show that dogs can easily be tempted to take bait without exposing endemic animals to any risk. A few years after my research project, Galapagos National Parks' staff successfully removed all feral dogs from Isabela.

I have learned much here in the Galapagos, by being in an environment that is virtually without natural predators, and this after I have become used to the Serengeti with its 25-odd species of carnivores. Being without such enemies leaves an ecosystem tragically vulnerable. I am left with even more admiration for Africa, for the Serengeti, and the Ngorongoro Crater.

The implications of my Serengeti dreams, my thinking about predators while sitting on a *kopje* on the grassland plains, have reached the other side of the globe. At times, nostalgia overtakes: even when watching those Galapagos dogs, my thoughts are still with dear, good Solomon, as I would hug him, sitting next to me in the Land Rover.

CHAPTER 17

Badgering in Britain

Aggressive male Eurasian badger.

C LANS OF HYENAS across the endless plains of Serengeti, packs of dogs hunting below the volcanoes of the Galapagos ... There is fabulous life in the beauty of those landscapes, full of excitements. But what about the quiet woods in Britain? Walking somewhere in England at night, between dark trees and dense undergrowth, I hear a robin singing and the occasional call of an owl. It is more difficult to see wildlife, even though there are animals about.

Leaving the Serengeti I find tough, but needs must. Suddenly things conspire against our wonderful existence in Seronera. My job at the Serengeti Research Institute is being 'Africanized', our young daughter needs a doctor more often than the Serengeti can provide, and back in Oxford a job beckons at Niko Tinbergen's research group. After all these years I have to leave, and I am back again in Britain.

Yet I am fortunate. I still have opportunities for research on animals in many other marvellous parts of the world, from England to Scotland and Shetland, to Thailand and Australia, from Alaska to the Brazilian Pantanal. Fabulous places. Masses of possibilities, but inevitably after Africa, wherever I am I have the Serengeti in the back of my mind and, as I felt in the Galapagos, Solomon seems to be looking over my shoulder. I often think of the hyenas, the vultures, the big Serengeti skies and horizons. The animals there taught me about their life and death, they directed my research on their behaviour, their social life and ecology. More personally, they caused changes in my feelings about wild animals, even about pets. The Serengeti taught me how to be objective about behaviour, about what animals do to each other, even about what we call cruelty. The Serengeti taught me that animals die violently. That, too, is a legacy from those huge, grassland plains, and it is with me, always.

In my new research project in Britain I am obsessed by European badgers; there is something mystifying about them. After many years in the field with other carnivores, with hyenas, wild dogs, lions, cheetah and others, I learned that many of them live their social lives as co-operatives, and in their societies help each other to hunt. European badgers also live in groups, in large burrows called setts. But weirdly, here among the huge old oak trees in the ancient Wytham Woods on a hill overlooking Oxford, these animals seem to be feeding almost only on insects, worms and wheat. There are no baying packs of badgers pursuing deer through the forests, no co-operative forces, no complicated calls or obvious behaviour displays. Why on earth, then, should they live in large groups, occupying these elaborate setts? I start to watch them, I start badgering.

The endless, majestic savannah of the Serengeti, the crater floor of Ngorongoro, are replaced for me by pretty English farmland, rolling hills with monumental trees, woodland groves, by cattle grazing the lush pastures. Old oak woodlands are crossed by small tracks, birds sing in abundance. And there are fences.

Perched high up on a branch in a magnificent oak tree with my wife Jane and a friend of hers, I am showing the two girls how one watches wildlife. Below, close to the tree, is a huge badger sett, it is evening, and we are awaiting the emergence of its occupants. I am the authority with years of experience of watching animals. I have told the girls to keep absolutely quiet, not to move whatever happens, to ignore the buzzing mosquitoes. It gets darker and darker, and we are staring at all those holes below us. Nothing comes.

I tense, and – triumph. The tip of a nose shows in one of the entrances, a bit of white, a badger will surely follow. Then, disaster. In my nose an explosive sneeze builds up, there is no way to stop it, and I suffer a massive outburst. It is the end of any chance of seeing a badger that night. The girls giggle, and my reputation is in tatters.

Yet, this is the beginning of my serious involvement with badgers. I want to understand their social life, find out what makes them tick, why there are so many of them in some places and fewer elsewhere. The animal is pretty well known to science and naturalists, and as a new arrival in this country where I am looking at one of its most popular animals, I take for granted that the locals are profoundly knowledgeable about the badgers' ways of life. I read a well-known naturalist's book about badgers, about the animals' friendly coexistence in large groups, about their social visits to neighbouring setts, about their hygienic dung-pits and their habits of feeding on just about everything. All this is existing wisdom among badger lovers. Yet sceptically, I feel that there is something wrong about that existing friendly wisdom. Life isn't like that. The answer to my unease lies in watching the animals, in following them wherever they go, and not just at the entrances of their holes.

Easily said, but watching badgers after they leave the security of their setts is difficult. They are nocturnal and quite shy when they are on the move. Things happen in darkness, either at night or underground. To do anything useful with badgers I need help, so I bring in some technology, and my fascination with these animals develops into a proper research project from the Oxford Zoology Department.

I find badgers are quite easy to catch in large box-traps or stopped snares, so I soon have small radio transmitters on some of the animals. They enable me to find badgers in the dark from a distance, wherever they are foraging in the woods and fields of Wytham. I also fit a small glow-light onto the radio-collar on a badgers' back. It doesn't seem to affect the animal, but it is a beacon to me, as eager observer in the pitch-black forests.

On a dark and rather damp night I am standing at the edge of a field in Wytham, keeping absolutely still, but aware of a slight wind in my face. Only yards away from me, the little, greenish glow-light moves backwards and forwards, staying in one small area. I don't dare risk frightening the badger by using a torch to see what is happening, so all I know is that it is there. Chomping noises tell me that it is feeding. After about half an hour the animal wanders off, still unaware of my presence, the radio signal disappearing. Instead of following it, I wait a bit, then use my torch to check on the places where the badger was foraging some minutes earlier. I find grass, but otherwise nothing. No signs of digging, no scratch marks, nothing. Later with my radio receiver, I find the animal back again in its sett, its den – it is the end of my observations for the time. Following it at night through the dense forest proves to be very frustrating, even with technology, despite my radio tracking. I am up against it. What does a badger do, and why? Is there some kind of badger organization? At this point the Serengeti comes back to me, with help from my hyenas in a rather macabre role.

Walking around in those woods near Oxford and following the many game-trails left by badgers in the undergrowth, I cannot possibly miss the many badger latrines that appear to be near the setts, near paths and roads, near fences, and all over the woods. There are scores of them, very conspicuous groups of small pits dug by the badgers and filled with their dung, with signs of a lot of scratching around them. Badger dung is large – some scats are almost like small cow-pats – with a curiously sweet, unique badger smell.

In the Serengeti, I analysed a lot of hyena dung and found tiny, coloured glass beads in some of the scats, beads as worn by Masai women. Sometimes there were many curly black human hairs as well as the glass beads. If hyenas can eat them, perhaps I can get badgers to eat coloured beads by putting out different colours in bait at each badger sett, so I could then find beads in badger latrines and map where they came from. It could tell me about where badgers move through the woods, more efficiently than from radio tracking.

Here in Europe little glass beads prove difficult to get, but I acquire some from Kenya. To persuade badgers to collaborate, the beads need to be used in an attractive bait near the entrances of the big badger setts. I try all kinds of thing – carrots, apples, wheat – in vain, the badgers turn up their noses. By chance, at the bird-feeding table at our home just outside Oxford, badgers also come every so often for a nocturnal snack. I notice there that they love peanuts more than anything; they make us feel that they will commit any crime for a peanut. To my delight, the day after I put peanuts out near the badger setts in Wytham Woods, bits of nut turn up in the badger latrines, and with it the odd glass bead that I stuck on to the peanuts with sticky molasses.

Later, coloured bits of plastic turn out to be at least as good as beads, and easier to get: with my assistant Peter Mallinson I cut up bright green and other plastic shopping bags in whites, reds and blues into tiny strips and shapes. Within days the badgers have decorated lots of their latrines in Wytham with bright colours and shapes. I can draw coloured lines on my wall maps of the badger area, connecting the setts where I put out colour-marked peanut bait with latrine sites where the badgers deposit my generous gifts.

It results in a striking map of the badger territories in Wytham Woods. Many of the latrines are exactly on territorial boundaries, and in the woods I find clear badger paths between them, showing borders. The result of this exercise is a pretty accurate size for each home range or territory. Every one of these has several badger setts in it. Later, badgers with radio transmitters show that they do, indeed, regularly visit several different setts, just as common lore had it – but almost only within their own territory. Outside that area, outside that home range, is hostile ground. Within each home range lives a group of badgers of a dozen or more animals, a *clan*.

Late one night I am following a male with a radio transmitter from one of my known clans, on a boundary path between latrines. In the pitch dark he

Female badger and cub.

passes and ignores another radioed male badger which I know to be from the neighbouring clan; it is feeding some 50 metres away, in his own territory. Standing between the trees and bushes, from the signals I know exactly where they both are. Then, unexpectedly when it is some 100 metres further, my animal turns into what I know to be the neighbouring clan range, and it starts to sniff around. Twelve minutes later he happens to walk upwind of the other badger I had noticed earlier.

That is his undoing. The other animal is on to him in a flash, and my badger turns and runs. The defender quickly overtakes him, and bites him hard amid loud, keckering noises. My badger escapes to his own ground, followed by his pursuer. There, 'my' animal turns on to his tormentor, chasing him back across the border line and well into the other's own patch. After which the tide turns again, and the chase reverses – but this time the two enemies race straight towards me. In the dark they almost hit my legs, giving me goose pimples and the badgers themselves the fright of their lives, scattering noisily towards their own setts. It is the end of my observations for that night. Incidents such as this cause the horrendous wounds one often sees on the rump of some badgers. These sweet, cuddly-looking animals turn out to be vicious fighters.

Somehow one can accept why hyenas, leopards, dogs or lions are dangerous fighters even against people: it goes with their lifestyle, with their kind of prey. But badgers? These friendly, small animals that roam the forests of our British countryside? Living in groups as they do, what strange method of hunting do they go for? More information is needed, more fieldwork. And just then comes help from an unexpected source.

Here in Oxford, my friend, boss and mentor Niko Tinbergen, the great animal behaviour scientist, receives the Nobel Prize for his pioneering work. Within days, and while we are still celebrating, he gives me one-third of his prize money to spend as I like, and I am now able to buy the latest, state-of-the-art night scope.

The night binoculars, which I call the 'hot eye', are sensitive to infrared light, and there is a large infrared searchlight attached to it. I can sit quietly at the corner of a field, in the pitch dark, and watch every move and every detail of a badger, even when it is more than 100 metres away. At night, even without any moon I can see behaviour in detail.

In a typical English, bleak, damp and dark night the radio-badger walks slowly, searching nose down, close to the ground just in front of its feet. No digging, no scratching, just walking quietly. Quite fast and without much effort, it grabs a large earthworm from the surface just to its left, and pulls. The worm is still attached with its tail end, its tiny bristles clinging to its burrow. But quietly, the badger pulls without snatching, takes the strain, and after a few seconds the worm lets go. A few smacking noises, and that is it. The badger walks on a few steps, turns slightly, and the same thing happens. Worm after worm, I count 18 of them within a quarter of an hour. The badger's art is to pull the worm slowly, not too fast, so it does not break.

After 20 minutes the badger wanders away across the pasture, while I get up and follow carefully. A 100 metres further it starts 'hunting' again, as if it were eating spaghetti. There is a strange kind of excitement in being able to follow an

Radio-collared badger searching for earthworms.

animal so closely, in the dark, with the continuous threat hanging over me that my quarry may detect me. That would mean an immediate hasty departure on the badger's part, and usually the end of observation for that night. The trouble is that I cannot move fast when the animal is close to me, because that would create a lot of noise (and even a few rustling leaves will spook a badger). I just have to stay immobile, even when I see the animal coming straight at me, or moving perilously near to coming downwind, when it would smell me and be off.

It sounds easy. But many is the time, during black nights when I am following an animal carefully on foot across the fields in the woods, that the world seems to conspire against me getting anything interesting out of it. I am using only a tiny torch to see where I am going without disturbing the badger. Cattle and especially horses in the fields are a nightmare, as they follow me around in the dark at uncomfortably close quarters, making an unholy racket in the pastures, which of course disturbs 'my' badger. The livestock rollicks around and tries to push or kick me, often setting my pulse going faster than is pleasant. I seem unable to reason with these huge beasts.

The badgers find lots of earthworms. Large ones, known as 'night crawlers', stand out especially. What astonishes me is the sheer quantity of them, a huge bulk of biomass.

As a rule of thumb, it is reckoned that a farmer here can count on grazing a number of cattle in a field, a number similar to the biomass of earthworms below the surface. To get some idea of earthworm numbers, I spend many days extracting them with a red watering can of diluted formalin, anywhere on pastures, in the woods, or in ploughed fields. And I find masses of these night crawlers (incidentally, during World War II British scientists did some chemical analyses and worked out that the protein of earthworms is like that of prime beefsteak, and would be excellent for people; alas, the idea was never taken up). In my own sampling, I commonly find 30 or more large earthworms per square metre of pasture in many areas in and around Wytham Woods. Each worm weighs some five grams – that makes about one kilogram of earthworms for every six or seven square metres.

This abundance is what the badger is after – these night crawlers may live deep underground, but they come to the surface during humid nights, a ready prey for the discerning predator. And after studying badger dung, masses of it, I now know that my Wytham badgers live mostly off earthworms. I know how they catch them on the surface, in patches mostly in short-grass pastures, the badgers often feeding in close proximity to grazing cattle.

The clan territories have boundaries around feeding areas in pastures, around earthworm patches, some of the borders coinciding with fences. It looks as if the size of the clan, the number of badgers living in it, is dependent on what their territory has to offer, on the biomass of earthworms. Probably, the territory protects the feeding patches, at all times of year. If these patches are far apart, the territory has to be large. If the patches are rich, with many earthworms, many badgers can use them at any one time without competing, and the clan will be large with many individuals. I think it is a gratifyingly simple explanation for the system of clans, for the sizes of their territories and the numbers of their members.

Being from a social species myself, I am always interested in who sleeps with whom, and the badgers' domestic arrangements give me nice social details of badger life. And as a voyeur following my radio-badgers, picking up signals from their underground sleeping chambers made me realize that there are rich pickings there. First, I notice that the older and larger, dominant animals, are almost always spending their daytime in the large main setts, the ones with many entrances, and they are sleeping with several others. In contrast, the younger or smaller, sometimes rather scrawny badgers are more often alone in smaller holes further away, in the 'outlier' setts, with only a single entrance, that are also difficult to find. It looks as if weaker (and perhaps diseased) badgers stay away from the main groups. They live in the badger slums.

After a few years at a critical time, my badger project is rudely interrupted. Niko Tinbergen, the maestro, retires. With that, his Oxford research group also comes to an end. I have to find a new job so, alas, my research in Wytham Woods draws to a close. It results in a move to the north of Scotland. But fortunately and almost inevitably, I soon get involved again with badgers there, looking at them in many fascinating and beautiful places, enabling me to compare between badgers in different environments.

Watching the animals in those ancient oak woods of Wytham was wonderful. It had all the essential ingredients for what I need in fieldwork: the rich tapestry of the largely undisturbed beautiful woods, with a high density of badgers and their setts. Here in Scotland I am far from the fertile woods and fields of southern England, in wilder landscapes, in huge, wide glens and moorlands, with birch forests and pine plantations, with herds of sheep as well as cattle, with fewer people and less traffic. Badgers are much thinner on the ground than down in the south. But there are interesting new research opportunities with the badgers here. I work in an ecological institute in the north-east of the country, with a large area of land around it that provides a new and excellent chance to look at the animals from quite different angles.

New and different ways are not always painless, though. In my new institute I can keep captive badgers (obtained from a badger keeper in England) in a large enclosure of wild birch woods near the institute building in the depths of Aberdeenshire, and they give me a new window into their life. It also causes me to be attacked by an animal as I have never been attacked by any before, or since.

Midnight, in a good old Scottish drizzle. Tim, a year-old male badger, walks a few metres ahead of me. Armed with a large torch, I am keenly interested in what he does, and exactly how he finds his food in the old birch wood enclosure. He finds the odd worm, and digs for some leatherjackets. He is tame and used to me, bottle-reared by myself, now living semi-wild.

He stops, and vigorously scrapes the ground with his hind legs. It is a clear social behaviour signal – I should have known it, but to my shame I don't. I ignore it to my regret. A few paces further on, I again see that hind leg scratching. He is telling me something – then Tim turns around, and flings himself at me, something that has never happened before. He growls, bites my legs and bites again, and I am painfully aware that my rubber boots are not much of a protection as his teeth go right through. Dancing around in the rain, in the pitch dark trying to avoid him, I slip in the mud and fall, and the badger pounces on top of me. My large bulky torch is the only weapon to hand. With it I bash the animal without mercy and as hard as I can – with the only result that the torch explodes into bits. I am in the dark, in the mud and rain, with a furious badger upon me, his weight feeling like a ton, and he is biting me everywhere.

Twice, from my muddy, prostrate and desperate position I manage to lift the badger up and throw him away from me, but within seconds he is back again, renewing his attack. My screams are heard by my old friend Charlie, the caretaker of the institute building nearby, who still happens to be there at

A sporran in Scotland.

midnight, cleaning. He rushes out with a light, scales the fence of the enclosure, grabs a stick and with it in the dark he starts belabouring what he thinks is the badger. More often than not, it is me. Tim now attacks Charlie, giving me the chance to get on my feet again, Charlie's legs now taking the brunt of the biting. Again I manage to pick up the animal and throw it. Together, Charlie and I somehow manage to reach the fence again and get over it, away from those ghastly jaws. There is blood all over the place: we both have deep gashes in our legs. Hours after midnight a doctor stitches us up again.

Since that occurrence, I hear about several other, bad attacks by semi-tame badgers on people, some even resulting in prolonged hospitalization. Charlie and I were lucky, but years later we both still have the scars to prove it. Badgers are not animals one should try to tame, they are too dangerous. One problem is that if a badger is aggressive, this is not immediately obvious as in a dog or a cat, but one just sees that vigorous scratching with its hind legs. In an attack the furious animal is never put off by whatever punishment one metes out to it. In my nocturnal fight, however hard I hit it, it just comes back for more.

After this event Tim, one of several research badgers in the enclosure near the institute, becomes totally unmanageable. This is especially evident when any of the females in the group are in oestrus. He comes for me as soon as he sees or smells me, and I don't dare let anyone else look after him. In the end, reluctantly on my part, he goes to a zoo, in Dundee, which is keen to have a vigorous badger for display, and they think they can cope with him. With thoughts of Solomon, I reflect that, again, my 'taming' of a wild animal has gone wrong. Months later, Tim the badger seriously attacks a zookeeper and, if that is not enough, in the middle of the day he escapes among the public and mauls a young boy. Sadly but inevitably, he has to be put down. Some badgers are much more aggressive than people give them credit for.

In Scotland Tim and the other badgers, in my home-bred colony of up to eight animals, teach me a great deal about their social life. I can even look

Female badger carrying food to her cub.

inside their sett deep underground, where the badgers sleep behind glass but with free access to their wild range.

The dominant male is usually ready to maul any other male in sight, including his own male offspring within the same clan. The matriarch, One-lug, a fat old dominant female who is at loggerheads with all the other females, kills cubs that are not her own, including her own 'grand-cubs', offspring of her daughters. She fights her sister, as well as the eldest daughter present. Because of that, every year I have to remove the eldest daughter in the latest litter, after they become independent, in order to prevent them from being slaughtered by One-lug. I release them into the wooded farm country nearby.

In our colony and until the time of independence when cubs start to forage for themselves, One-lug looks after them very well. She even carries morsels of food to the cubs, such as dead day-old chicks that I often give the badgers. I had never seen this before in the wild. One-lug is also the badger who does most of the nest-building, gathering straw and bracken, dragging bundles of it backwards over long distances, clutching the gatherings behind her forelegs.

I am intrigued by these goings on and the individual quirks of the animals. One fascinating, though not totally unexpected, finding from my semi-captive clan of badgers is that any aggression between the individuals builds up to a horrible crescendo if their food gets scarce. Immediately when I reduce their daily rations somewhat, fighting becomes more serious, despite them living in a large woodland enclosure in an almost natural environment. There is not just fighting over food, but there are attacks on the 'under-badgers' at any of their meetings. Give them more food, and the nastiness largely stops. If I do not remove badgers that are being picked on, they will surely be killed – and in the wild, they would need to emigrate. This, I think, is part of their main natural population regulation, although here in my captivity set-up, of course, they do not have access to normal, natural resources.

Keenly interested in the way in which badgers find their food, when I watch the animals in the wild I get ideas about what they can and cannot do, and what constrains them. One way to test my ideas is in the enclosure. Observing from outside the fence, with me out of sight of the badgers, I have put down several grass swards inside, patches of a few square metres of lawn turf that differ from each other only in the length of grass. A few large earthworms are tied and pegged down in each patch (though tying down earthworms is a job I do not wish on my worst enemy), and I can simply time the badgers' efforts to catch the worms with my stopwatch.

I can clearly see now how long grass hampers the badgers, making it difficult for them to keep their nose to the ground, and an earthworm that is not tied down would have ample time to withdraw into its burrow and escape because of the racket made by the badger in the vegetation. In my experimental short grass (such as grass tightly grazed by cattle), badgers catch a worm on average in five to ten seconds, but when grass is 20 centimetres long or more, this time increases to 30–50 seconds. I think it is one of the reasons why badgers forage in pasture that is grazed by cattle.

Much more exciting, Scotland provides some excellent wild areas for badgers. Ardnish, on the west coast of Scotland, is a wonderful world. A remote peninsula several miles across, without any people, it is steep and mountainous, open moorland with patches of oak and birch, and fabulous views over the sea with small islands. With my old friend Ray Hewson, who is studying the foxes there, I stay in a primitive abandoned house surrounded by sheep, red deer, otters, ravens, eagles, seals and badgers. What more can I want? Badgers are not many, but that is exactly what makes them interesting here – why are there so few?

The place feels like an island rather than a peninsula. A small, tricky footpath through high hills connects with civilization, but to reach our bothy at the end it is easier to cross the sea loch with a little boat, despite the often atrocious weather conditions. There are ruins of old crofts, long abandoned, and the walls of a former school. The Gaelic names of many of the sites bear witness to a world long gone. There are remnants of stony footpaths where little bare feet used to tread to school, small beaches where fishing boats used to be hauled out. They are leftovers from a tough, poor, exacting life of crofting people in a wild, wild place, haunted by ravens, eagles and gulls.

There are no cars here, so everything as to be done on foot. Following Ardnish badgers at night is pretty hair-raising, even with my radio-tracking equipment and infrared binoculars, the 'hot eye'. The slopes are steep and rocky, much of them covered in bracken; I always seem to end up in the most difficult bogs, and the narrow sheep paths are made by animals more sure-footed than I am. During nights with little wind, the swarms of midges are horrendous. Dawn comes as a great relief, when the badger retires into its sett and I can subside on a stone along the shore, watching otters just for a change.

Miles away from anywhere on Ardnish, miles even from any sign of previous habitation in this overpowering landscape, is one of my badger setts. It is clearly an old one though small by badger standards, well used, in a slope of a sandy patch quite far inland. One morning, when I am passing there accompanied by the calls of ravens and gulls, I catch a glimpse of something unusual outside one of the entrances. There are clumps of old bedding, of grass and bracken cleared by badgers from their tunnels, as they do every so often. In the discarded badger bedding I find a small, perfectly executed bronze bust of King George VI, of a size somewhat larger than a badger's head.

Many are the drams of whisky downed in our bothy while we guess at the origin of 'George', now safely ensconced above the fire place. Who could ever have put his bust into a badger sett, so far from civilization? Ray is convinced of something more sinister, and declares that 'George has the baleful eye'. For that reason, he thinks, the bust has been safely disposed of by a previous owner. Here, when he is above our fire place, the baleful eye is only to be appeased by a daily dram of malt whisky in his honour. Some years later, George is violently attacked with an axe by a mentally unstable Dutch zoology student.

I remind myself that the reason for me being in Ardnish, in this magic place with fantastic views over the sea and the islands, is badgers. Befitting the ambience, the set-up is quite different from what I am used to with badgers elsewhere. Unlike the badgers I know down south in England, here they space themselves widely, their setts kilometres apart with only one or two badgers per sett. Sometimes the Ardnish badgers share their sett with an otter. When they feed, they take a sheep carcase apart as if they plan to carpet an entire

Badger habitat on the Scottish west coast: Ardnish.

glade with the wool, making a terrific mess and sharing it with ravens. When I watch from the top of a rock at a distance, the ravens in turn need to stand back to give way to a golden eagle, gliding in spectacularly from up high. Conspicuously, the Ardnish badgers fill themselves from sheep carcases, with beetles, small rodents, or with rowan berries in the autumn. Their main food, however, as is obvious from their dung, is … earthworms.

The small patches of short grass on the Ardnish slopes, as well as the more extensive coastal greens, are intensively grazed by the sheep. They are therefore covered with lots of sheep dung, enabling many small worms to feed here. And interestingly, when one year the owner of the land decides to stop keeping sheep here and encourages red deer instead, it means the end of grazing and short grass, and the end of the badger story. It is almost a natural experiment: because of the different grazing, the short grassy patches are swallowed up by tall and rank vegetation, worms are almost impossible to find and the badgers almost disappear from the peninsula. Their setts are abandoned.

The disappearance of the badgers is the end of my Ardnish episode. Alas, the end of a time when Ray and I battle to the mainland across the wild sea loch in our little boat, in order to avoid the long and difficult walk across the hills. Sometimes we get soaking wet, sometimes we come close to disaster in the wild weather. It is the end of our foraging for large oysters and clams at very low tides, it is the end also for our many early-morning otter watches when, huddled in a sleeping bag, one of us sits at the window of our bothy and shouts out timing observations on diving otters, while the other writes them down. Ardnish, always intensely beautiful, provided many insights and delights, as well as being an awe-inspiring place.

CHAPTER 18

Olfactory delights and olives

Part of a clan of European badgers.

ROTHIEMURCHUS, my next badger study area in the central Scottish Highlands with its splendid Scottish ring, does not have the same glamour as Ardnish with its west coast landscape. Yet it is beautiful, with its birch slopes, with Loch Pityoulish, the River Spey and its sheep-dotted pastures. Badgers are numerous. They dig themselves large setts with a dozen entrances or more, much larger dens than anything the Ardnish animals produce.

One boiling hot afternoon in June, after collecting badger scats in Rothiemurchus to study in the lab, I strip off and take a dip in the loch, swimming quite far out. Lying on my back in the water. I get an overview of the main badger sett that I am now studying here on the slope above the loch. I can see its entire territory. I know where this sett's feeding patches are, I know its clearly defined boundaries, having radio-tracked several badgers from there in previous nights.

Floating in the cool water I can take it all in, and I am getting ideas. Foraging out in the fields and in the birch woods, I find the badgers are as solitary as can be, so at least some secrets of their social life must lie in the sett. Why do these animals live in communities like that? Do they help each other? Living in quite large groups, with other, sometimes horribly aggressive and individually not distinctive badgers (especially not in the dark) – how do they know who warrants aggression?

Later in the day I sit against an old tree and watch the sett after the sun goes down. I wait for things to happen, while being pestered by midges. Tired after quite a bit of slogging on the hills, I am thinking how nice it would be to be at home with the children. Then, once a late osprey flies past, I am again beginning to feel glad to be alive.

A badger pops up from an entrance just below me, and I am keeping mouse-still. I can see that it carries a radio-collar, though my receiver gives me no signal. At the moment there is only one badger in this entire area that carries a non-functional radio: his nickname is Hitler. He is especially interesting to me: catching him two years ago in the autumn as a smallish cub born that year, I tattoo him, and catch him again half a year later in the very same clan territory, different from the one in which I am sitting. Once he has a radio-collar I notice that he is unusually aggressive to other badgers. And he has only one testicle – hence the nickname. Now, a year later, he is here, in the territory of the neighbouring clan! He has changed allegiance.

I watch him poking about the entrances. When another badger emerges and shakes itself with sand flying all over the place, it walks up to Hitler. After some hesitation the two sniff each other, and then I see a wonderfully interesting piece of behaviour. One of them turns around, both lift their tails, then they back into each other, bum to bum. A thorough rub, before they walk off in opposite directions. I had seen this before, but now I realize the possible

Scottish badger den, or sett.

significance of this bum-press: it means integration. Hitler is being given a passport into his new clan.

Our Eurasian badgers have an enormous gland just under their tail, a large pocket above their anus. It is full of a white, greasy secretion, which to me smells like soap. My colleague Martyn Gorman has demonstrated that in that secretion each badger produces its own personal smell, a scent profile with more than 20 different 'scent peaks'. The relative heights of these peaks produce the individually different aromas. Badgers belonging to the same clan are more similar in their smell than badgers from other clans. And the smells are produced by bacteria living in the badgers' greasy secretion.

So what I am seeing on the badger sett above the glistering loch between the birches, is Hitler, the recent new arrival, mixing his bacterial flora with that of another badger in his new clan. As far as smell is concerned, he is now one of the clan. It seems a perfect way for badgers to achieve a large, happy and smelly family out at night in the dark, and in the deep black tunnels of the sett.

The question now is, what do they use it for, this social cohesion? Do the members of a clan help each other, somehow? For me it is one of the most important problems and rather an enigma.

None of the badgers is aware of me while I am still watching the animals on their sett, in the growing darkness. More of them emerge, and three cubs begin to play and scratch an old tree-stump. The torture by midges is getting too frightful, they are stronger than me and I just have to wipe a crowd of them out of my eyes and hair. One badger, right at the back of the group on the sett, happens to look around at that very time, sees my movements, gets a fright and without any preamble dashes noisily down the nearest entrance. Amazingly, there is no overall alarm, and the other badgers just continue with what they are doing.

Compare these badgers with the hyenas back in the Serengeti, which I often watched similarly, a group of them playing about on their den. A movement like my foolish mistake of wiping off the midges, if it were spotted by one hyena,

Badger male scent-marking with sub-caudal gland.

would have a much greater effect – there would have been an alarm call, their typical soft, deep and rapidly repeated grunt, and the entire crowd of hyenas would have dashed off immediately, out of harm's way. Of course, that is what co-operation among animals is about, they warn each other. But these badgers are just not in the same social league as the hyenas – for badgers there is no alarm call, it is everyone for himself. It almost suggests that they are right at the bottom of the ladder of social evolution, they do not even warn each other.

Perhaps the only way in which badgers really help each other is right here in front of me, with the construction and maintenance of their huge setts, of the 30 or more large holes with sometimes literally miles of tunnel inside. In the setts, badgers may sleep happily curled up against each other, maintaining warmth during their long period of inactivity in winter. Their 'hibernation' is more or less compulsory, enforced by their dependence on the seasonality of earthworms and insects on the surface.

It is not everywhere that these animals can have the comfort of communal setts and sleeping-huddles. They can only afford this if the food supply allows it, such as the masses of worms in British pastures. Badgers in the wild, remote hills of Ardnish, where eagles fly over the poor soils, are reduced to an enforced solitary existence. It all comes back to food again, one of the most important environmental pressures on society anywhere.

I think of British badgers as animals of pastures and of rich woodlands. But there are places elsewhere where badgers live in much rougher locations, in the mountains. Ardnish is one such, and in Italy I found others.

Naturalists with fertile imaginations in the seventeenth century reported that badgers' left legs were longer than their right ones – this was seen as an admirable detail from their Creator that enabled easy walking clockwise along steep mountain slopes. When I come to see badgers going up and down the very sharp slopes of Monte Baldo near Lake Garda in northern Italy, I cannot think why anyone would have thought that these animals needed such an amazingly nonsensical adaptation. Going up the mountain slopes at a great clip, any badger leaves me way behind and panting, battling the dense undergrowth of the macchia. Badger paths here run up and down through some 800 metres difference in altitude. They leave me feeling as if they are going almost straight up, and I can only wonder at the enormous amount of energy that is wasted by these Italian animals, carting their heavy badger bulk all the way up, and down again, and often.

On the face of it, there seems to be a world of difference between the badgers I know in Britain, and the animals here in the Italian mountains. The Monte Baldo setts are small, with no more than two or three entrances, and in many the badgers live on their own. Even more interesting is their different food. I am used to highly specialized worm-devouring badgers foraging on short grass around cattle or sheep, but on Monte Baldo, quite appropriately, they know the delights of olives, a good Italian product. And the animals make this abundantly clear: every bit of badger scat in the latrines long the slopes I find is brimful of olive stones.

Out on the mountains I cannot but love this magnificent country, with its spectacular views over the lake and the distant Alps, and its fabulously rich

Badger habitat on Monte Baldo, Italy – the upper parts.

flora of shrubs and flowers. Our friend Lil de Kock invited me here with the family, to stay in her luxurious villa surrounded by olive groves just above Lake Garda. Lil is an excellent naturalist who used to live close to our house in Scotland near Aberdeen, after her hair-raising escape from Nazi Germany in the 1930s. She knows Monte Baldo like the proverbial back of her hand. As a botanist she specializes in orchids, and notices the many badgers among her own olive trees. She realizes there is something unusual, and challenges me to find out what is happening.

Low on the mountain slopes I am walking through age-old olive groves on stony terraces, beautifully contorted old trees, well kept and pruned. The slopes are studded with a few old farm buildings, with cherry trees, vines and figs. The olive groves are well patronized by the badgers, which is abundantly obvious from the first badger latrine that I find next to a narrow path. The pasture in the olive groves looks lush. Could it also be an earthworm paradise?

In the beginning I pooh-poohed Lil's idea that in Italy a badger does not eat earthworms, but olives – but I soon find that she is largely right. Earthworms are very few here (perhaps the seasonal droughts kill them), but there are masses of olives on and underneath the trees, and most importantly, olives are there at almost any time of year. Badgers also eat the many other fruits available, but they are much more seasonal. In these scenes of Italian exuberance, it is the oily nutrition of the olive that badgers are after.

I am quite prepared to agree with Lil that we can call the badgers here olive specialists. But I soon find that there is more to the story. Panting up the mountains, following badger paths and finally getting above the treeline, I am still in badger country on the open mountain meadows with their breath-taking views. This is high ground, some 1,000 metres up, and quite inaccessible under deep layers of snow in autumn, winter and early spring. This is where the cattle are brought to graze in summer. But this is also where there are earthworms, lots of them in just a few small patches here and there, but absent over large areas elsewhere. And as expected, in the badger latrines here, the badger scats do contain the minuscule hairs (chaetae) of earthworms, which I can identify under the microscope at home in Scotland.

But apart from earthworms, even high up the mountain these same badger scats contain olives, lots of them, collected way down below the high pastures, with long, hard hikes up and down the mountain. Also, along these slopes badgers appear to eat insects, slugs and even reptiles such as lizards and snakes, as they do everywhere else. It does not diminish the finding that generally and right throughout the year, Monte Baldo badgers eat mostly olives. So while they may eat almost anything, their food is dominated by just one item, just as with the badgers in England and Scotland. That one item has a large influence on their social life, on their ranges and territories, and on their daily movements, even their group organization. They are specialists – here in northern Italy on olives, in Britain on earthworms, and their organization is adapted to this.

So extreme specialization is not something strange to badgers – be it olives, earthworms or, elsewhere in Europe, rabbits or cereals or another different prey. In Britain their diet is almost prepared for them in many places in the grazing pastures, by agriculture. It enables the animals' first step in their evolution as group-living carnivores.

Years later, a major international jump takes me to another kind of badger altogether. Smell comes into it again. I am retired now, and I finally get the chance to go and see an animal that I read about more than 20 years earlier. It was then described in a conservation journal by the naturalist Ian Grimwood as 'the ugliest mammal ever', a tiny badger much smaller than the Eurasian badger at home. I just had to find and see that Palawan stink badger. Aptly named, it occurs only on the Philippine island of Palawan and a couple of small islands nearby, and yes, if provoked it can produce a smell that would make an American skunk seem amateur. Locals call it the *pantot*.

It is hot, humid and sweaty. In Palawan, Jane and I are lying on our split bamboo bed, above a split-bamboo floor in a split-bamboo farm house, next to a small dirty glass window. I am dozing away while waiting for Frederico, the farmer, who will take me out at midnight to his rice paddies for a badger watch. A rooster calls throughout the night. A rustling noise alerts me to a large brown rat, half a metre from my face on the windowsill, silhouetted against the moon outside; a pig is grunting and bustling below the thin floor. Earlier, a hen tried to lay an egg in Jane's sponge bag.

Sleeping seems out of the question. I think of the long road we took to get to this small village of Aborlan on Palawan, by boat and bus through a magnificent lush scenery of high mountains, rice paddies and small farms. I think of the many people who directed and helped us, the clouds of dust around the crowded bus, the many new birds. Gradually I seem to drift off.

When the call comes I bounce up, leaving Jane nicely asleep on the split-bamboo beside me. Frederico and I walk along the moonlit track, just aware of mountains in the distance. It is still and quiet, there is a faint single noise from a distant owl, and when we get to the rice paddies the moon is barely sufficient to show the high ridges, the bunds, between the muddy rice fields. Walking between those paddies in the pitch dark, it takes quite a bit of balancing on the bunds to avoid splashing into the mud of the rice fields. Carefully, I just follow the experienced bare feet of Frederico. The rice plants are still small, and it should not be difficult to see if anything were moving about between them. But nothing shows in my torch beam. After an hour of balancing about, Frederico and I go our separate ways, to increase our chances.

It takes a bit of time before I hear a triumphant shout, '*Pantot!*' No attempt to be quiet, because there is no need: I find Frederico standing next to a small, rounded, black animal, with a very long snout and a whitish head. It is keeping still, facing us, its tiny beady eyes glaring. My first thought is that it looks rather like a hedgehog – same size, same reactions. But as I come closer and move around, trying to take some pictures, I can see why Grimwood was so impressed by its lack of beauty, especially when the animal keeps on turning its large, round and bare rear end towards me. I remember Tim the badger scratching the ground with his hind legs, and the lesson I learned from that.

I have heard enough about the Palawan stink badger here that I can see the warning sign, which tells me that if I came much closer, a squirt of foul liquid might come my way. The *pantot* has a reputation for accuracy as a sharpshooter. Frederico and I stand back a bit and fortunately, before long, it starts foraging again, silently in the wet paddy between the young rice shoots, digging here, walking a bit, and digging again. I notice its penny-size footprints in the mud, its claws like very small, Eurasian badger feet.

I am immensely pleased at catching up with this little thing, having come thousands of miles to see it. It is a totally different world from the Eurasian badgers I watch 'worming' in Scotland. Yet there is as a common factor the dependence of both badgers on some kind of agriculture that produces lots of prey. It is again the wild animal using the man-made landscape, adapting to new opportunities. Rice paddies, with their muddy surfaces mostly covered

Palawan stink badger, or *pantot*.

under shallow water, provide ample fare for the *pantot*, which eats small freshwater crabs and large insects such as mole-crickets. Each of these is quickly unearthed by the stink badger's long thin claws, and by the narrow snout poking into holes. Frederico tells me that he likes the *pantots* in his paddies, as the little crabs, crickets and others cause much damage to the rice.

The animal may be small, much smaller than the other species of badger I have watched, but it can certainly stand its ground as far as digging is concerned. After we watch it, the *pantot* walks up to one of the bunds between the rice paddies, starts digging and, within only a few minutes, it literally disappears into the hole it makes. In my torchlight, only its ugly, bare rear end shows occasionally. The performance is as impressive as that of the honey badger in the Kalahari on its sandy slopes. However different they are, both these species of badger move about in their tropical lands in a way very unlike that of the European badger, which always comes back to its large central den. Like the honey badger, the *pantot* often digs a new hole just for the day.

The little *pantot* is common here in the Philippines on this island of Palawan, and on tiny Busuanga nearby. Palawan is large and beautiful, its rice agriculture interspersed with high mountain areas of wild forests. The small badger fits perfectly into this partly man-made environment, people seem happy to have it around, and I have no worries for its future.

Otters. (Diana Brown)

CHAPTER 19

Shetland otters

Male otter along the Shetland coast.

WHILE LIVING IN SCOTLAND I just cannot avoid the otters. They first push themselves into my ken when Ray and I are staying in the old bothy at Ardnish, where I have come for the badgers. When waking there in the morning and looking over the sea loch from my sleeping bag, I may be thinking of my badger study, but otters come fishing just in front of me, while gulls call on the roof, ravens circle overhead and curlews burble. Otters are part of that wonderful and unforgettable country.

Once I have seen them there, somehow, otters seem to be everywhere in Scotland. Yet individuals are shy and elusive, and in most places surprisingly difficult to watch and follow. Near our house they live their lives at night, hidden in their watery environment of streams, lochs and reed beds, visiting our pond in darkness. Their signs are along many waters, but the animals themselves remain hidden. People tell me that there are places to see them easily, places such as Skye or Shetland, where otters are active by day along sea coasts with easy access. I realize that there I could make a start with finding out what they are like. Hopefully, in some easy place I can start a research project by instinctively getting the 'feel' of them. I really want to have a go at these animals, as they live in such wonderful places. They belong to the beautiful Scottish lochs and rivers, to wild coasts where they are surrounded by interesting environment and other fascinating animals. Theirs is a background of very rich tapestry, just what I need. At the time when I consider a project, not much is known about them, apart from the fact that there are serious problems with their survival and conservation everywhere in Europe.

It is autumn in Shetland, and I slowly make my way along a rocky shore not far from the bothy where I stay, early in the morning. For once the sea is calm, the tide is coming in, gulls are continuously overhead. I am on a routine walk, trying to get some more data on what otters are eating. It is sunny with a few busy birds, turnstones and redshanks poking around along the edge of the water. I detour, so as not to disturb them. Close to the rocks large beds of kelp wave on the surface of the sea, moving relentlessly, fronds pushing each other. There is space, horizon, skies with distant large birds. Shetland is as close to the Serengeti as I will ever get in Europe; it is different, but there is something similar in size and feel. Somewhere out there a seal will be swimming past, a small flock of eider ducks are cooing. It is wild nature, and I am made infinitely small. I walk with memories of hyenas and badgers, though around here my concern is otters.

A couple of hundred metres ahead I see what I am looking for, the by now familiar, small otter head, quietly floating along the smooth surface about 20 metres from the shore. I stop and keep still. Down it goes, out of sight, so I can progress a bit further without disturbing it, guessing how long the animal will be underwater. I stop and crouch; the otter pops up again. It is quietly scanning the surface, does not see me and is unaware of my presence. Fifteen seconds later it goes down again, giving me the chance to be quickly up and running forward, to try to get close, shielded by a large boulder. Panting, I crouch behind this, just before the otter pops up again, now much nearer. Lovely.

I am now able to record and time a good series of dives by the animal; I can see every bubble behind it. The otter is a female, and when she shows her throat

Shetland evening.

I can see the irregular yellowish-white patch, which almost all Shetland otters have, like a fingerprint. It is a good old friend, she even has a name: Wiebka.

Several of her dives are successful, and I am so close that I am able to identify the different fishes she comes up with – eelpouts, rocklings and butterfish, quite small ones that live at the bottom. They are all fishes that are active at night, but inactive and therefore easy to catch in daytime, hence the otters' daytime activity along sea coasts. I estimate how far offshore Wiebka is when she is diving, so from my contour map of the waters I can calculate the depth of catches. I time her dives with a stopwatch, and how long she takes to 'catch her breath' for the next dive. All good, basic and essential information.

After almost half an hour, she has had enough, having all that time fished in just one rather small patch of sea. She swims along the shore and lands quite close to where I am, still blissfully unaware that all her doings are being recorded for science. She walks up to the edge of the grassy peat a short way from the water, rolls and rubs herself, then curls up in the sun.

Half an hour goes by, and I just sit and wait. I watch the world, the birds, keeping half an eye on the otter. A gull circles overhead, some sheep come and stare at me. Then something exciting happens. Another otter approaches, swimming along the coast about ten metres from the shore. Even from here, quite a distance away, I can see that it is a large male, bigger than Wiebka, swimming in such a manner that I would have called 'strutting' if he had been walking. His tail floats along the surface behind him, his head held slightly further above the waterline than if it were a female. He is really conspicuous – a typical male, doing his round. I keep completely still, happy with having two otters in my world.

Swimming along the coast near the sleeping female, the male slows down and doubles back – I think he must have picked up a scent. The female is asleep, oblivious. Then quite unexpectedly, the big boy starts fishing, and I am beginning to suspect that I was wrong about him detecting Wiebka. After a few dives he comes up with a quite sizeable rockling fish, and lands it.

With the fish, he walks up to the sleeping female, who at first is obviously not best pleased to see him – she caterwauls loudly, and backs off. The male briefly walks round her, and drops the fish in front of her. The two sniff each other somewhat warily, then she takes the fish and eats. A couple of minutes later both are curled up, quite close together and peaceful. After another while the male walks back to the water and swims off again, disappearing out of sight.

Why should this ostensibly solitary animal be delivering a precious fish to the apparently none-too-grateful female? It is difficult not to be anthropomorphic, though I am reminded of my own courting days, arriving with a bunch of flowers at Jane's home. Perhaps this scene here has a similar explanation, putting the male in a favourable position. I saw the white identity-patch on his throat while he was fishing and I know him as M4, a long-term inhabitant of the coast of this entire peninsula.

Some six months later, during the next spring and not far away, I happen to be following that same male otter again. Now I at least get some answers to a few of my questions from the previous occasion, because again I see him meeting up with the very same female, Wiebka, both swimming. This time there is far less initial acrimony, and he mates with her. I feel a bond because perhaps, just as I thought, his courting present several months ago, his 'bunch of flowers', did get him somewhere. Otherwise, I notice that whenever otter males and females meet there is often aggression. It may well be that a strange, unknown male would have considerable difficulty getting close to a female, even when she is in oestrus. Just possibly, if she knows him and especially as a source of food, she may well relent more easily.

Seeing the two cavorting in the water and mating (which otters always do in the water), I realize again how complicated their social system is. Anywhere in Shetland, or in the rivers and lochs of mainland Scotland, one is always inclined to call otters solitary – but they aren't, really. They live and know each other in structured communities, but spend so little time together that anyone can be excused for classing them as solitary.

There is much going on between them that we are totally unaware of, for instance when the animals are communicating with scents. For otters that often means using spraints, the small pieces of excrement that they leave wherever they go, and which they produce 40 or 60 times a day or more. All that contains an abundance of information for other otters, as well as for any scientist interested in the animals. Otters may be difficult for us to see, but their spraints are conspicuous – in fact it is the presence of spraints that makes most people aware of otters.

Just walking along one of the wet, grassy and peaty shores here in Shetland, at any time I may come across one of these spraint places that tell a story about the animals. That day the wind is in my face, a cormorant is fishing, a tern calls sharply, there is a smell of kelp. In the soft, peaty grassland along the seashore is a hole, a tight tunnel going vertically down and filled with fresh, dark, peaty water. Around it the ground is worn and well trodden, grass well flattened by otters' rolling, greened by frequent urination, and there must be at least a dozen spraints. There are several such watery holes along the coast in this area, and it took me quite some time to work out what they meant. I have come to call them 'squeeze holes', bathing places where otters wash the salt out of their fur. These and other freshwater places are utterly essential to the animals, and they enable otters to exist along a sea coast. Fur is their only protection against the cold when they are fishing, a protection that falls by the wayside if it is not carefully managed. My own hair feels like a sticky mess after swimming in the sea and not washing out the salt. Swimming here for hours without an adequate barrier against the cold would make life pretty well impossible for a warm-blooded animal such as an otter. Seals have no such problem, with their thick layer of blubber. The spraints around the squeeze hole show how important it is, and they tell other otters that the squeeze hole is 'occupied', is being used by a resident and it is better to go elsewhere.

All the otters' dens here, their holts, even those unexpectedly high on a hilltop, have a compartment that serves as a bathroom, for the same purpose as the one in our own houses: to wash their skin and their fur. Some of these holts are huge, with complex tunnel systems just like badger setts on the mainland. Somewhere in the maze of tunnels there will be one or two bathrooms, with fresh, sometimes running, water from a small underground stream. Since

Otter holt in Shetland, with many spraints.

noticing the otters' need for fresh washing water in Shetland, I see it in many places, in ottery sites along any sea coast, anywhere, be it in Scotland, but even in Africa or America, and I know that a coast is otter friendly if fresh water is available in puddles, squeeze holes or in some other form. If there is no fresh washing water, then there are no otters.

The sharp 1960s decline in numbers of these very popular animals, of which so few were left here in Britain, was part of that silent spring caused by pollutants throughout Europe. Today DDT, dieldrin and others have been taken out of use, and otters are coming back or have come back, like many other birds and mammals that suffered. Even so, I feel that there aren't as many otters here as one might expect, even along these wild Shetland shores with all their fish. It is the reason for my main question. What is keeping their numbers down?

Here in Shetland that question should be soluble. Unlike the area near our house in mainland Scotland, or anywhere further south in Britain, otters are quite easy to see and study along the Shetland shores. They never disappeared during the pollution crisis, and where they live in the sea they are active in daytime, eminently visible. If I can get some solutions here along these sea coasts, I can use this understanding on the Scottish mainland and elsewhere, where things are a bit more difficult to observe, and perhaps it can help otter numbers along.

When I start putting sightings and observations of all my known Shetland animals on the map, it begins to show a system in what first appears to be chaos. Wiebka, with or without her cubs, is always along just one five-kilometre stretch of rocky shore, her home range, and in that home range she is most often along a short bit of coast, with some beaches and peaty bogs that I call

Shetland otter family sprainting (scent-marking).

her core area. Other females do more or less the same. They do use each other's home ranges, but not, or rarely, each other's core areas. The Shetland coasts are neatly divided up between these females, and I now also know that neighbours are often mothers and daughters, or sisters. Males move along much larger ranges of coast, each of them covering several female areas.

I am thinking of all this later, when crouched against a sheltering rock along the Shetland shore, overlooking the Sound between the islands of Mainland and Yell. But why do otters bother with these rather strict home ranges when there are so many fish in the sea, and they seem to be able to catch a fish just about anywhere? And as there are so many fish about, why aren't there many more otters feeding right here?

Where I am sitting I have to make do with just one otter in the field of vision of my binoculars. I am timing its dives, recording every detail I can pick up. Interestingly, in this large, wide expanse in front of me the animal dives repeatedly, again and again, in about the same place. It could go anywhere, but it doesn't. I wonder, because when I am catching fish in my small cage-traps, I have success almost anywhere all along the coast. With what I am seeing here I realize that otters catch fish mostly in small, restricted sites. And I find that these are places where they have access to the bottom without being bothered by the enormous forest of kelp. Otters do not catch fish just anywhere. I can see why when snorkelling, because for me, and perhaps also for an otter, swimming inside the tangled kelp forest is almost impossible. The fish are everywhere.

Walking later along the low Shetland shore, far from houses or people and keeping a close eye on the line between land and water to pick up any sign of otters, I catch a huge sigh from the sea, quite close. It is a pilot whale, surfacing some 30 metres away, breath-taking for me as well as for the animal itself. Its huge, silent size in the large seascape behind it puts my otter interest in perspective. The giant simultaneously disturbs my concentration and improves my sense of intense well-being, before it disappears again.

In the distance ahead along the shore, grey and common seals are on the rocks, with loud moaning calls from the greys, sounding somewhat like the rutting calls of red deer. There is small chance of seeing otters here, as they are never that keen on the grey seals, the aquatic bullies which every so often chase the otters. Though otters themselves do like teasing the common seals, occasionally nipping their back flippers or walking right up to their sleeping faces.

I walk along a hillside not too far from the sea, a place where I went yesterday without noticing anything in particular, apart from a small otter den – and today that same hole has a large sheet of dirty white plastic sticking out of it. It is the same nasty kind of plastic that annoys me so much on the tideline, with much other rubbish dumped in the sea. On the Shetland island of Unst is a large rubbish tip at the bottom of sea cliffs, close to a military base. I loathe those plastic tokens of human presence, but otters use them. Plastic sheets and fertilizer bags are dragged deep down into their holts, down into the sleeping chambers, just as otters drag bunches of seaweed, as bedding. It is their plastic furniture – I just hope that some conservationist is not going to demand that tideline plastic be protected, as otters use it. Maybe at some time the animals' furnishings will go back to nature, in the absence of plastic.

I take some pride in being invisible. When watching any mammal, especially the shyest and most suspicious ones, for me it is a *sine qua non* that it is unaware of my presence, that I melt away in the background. I will not interfere. I am heartless even when it comes to watching an antelope torn apart by predators, or a baby otter or a hyena mauled by an adult. I am only a witness. To be able to report, I feel it essential that I am not there, and that nature has to have its way regardless. But there are times when my objectivity is seriously challenged.

Wiebka, my favourite otter living along a shallow coast in Shetland, is probably not very old when I am watching her today, maybe three or four years. But she has a somewhat aged look about her, even when she is working hard at

Family resting on Fucus seaweed.

catching small fish between the seaweed, as she is doing now. It is late morning, grey and wind-still; I am curled up against a large boulder on the shore near the waterline, some 30 metres downwind from the animal.

Down she goes. I know it is four metres deep there, 20 seconds – briefly on the surface, down again, 16 seconds and up with a fish, a small, wriggling, bottom-living eelpout. She eats it while afloat, I can hear her chomping. I spend more than an hour watching and recording her there, with every moment bringing a new, exciting event, a new fish or crab or whatever.

Then everything changes. It starts with Wiebka catching a quite spectacular fish, the curiously named bullrout or father-lasher. It is thickset, well over 20 centimetres long, with a large mouth and spines on the back, its big, bright red belly strikingly adorned with white polka dots. Too large to be eaten while afloat, the otter takes it ashore. But in the process of being transported along the surface, one of the bullrout's enormous pectoral fins folds across Wiebka's face so the otter cannot see ahead, and she is blind as to where she is going. This strange assemblage heads straight for me, in my hiding place, and the otter, with her large fish, lands within two metres of my feet.

The flapping polka dots are dropped on the bladder-wrack, Wiebka looks up – and freezes. Clearly horrified by my appearance, she grabs the fish again and dives back into the sea, immediately out of sight – then reappears quite far away, closely hugging the coastline and difficult to spot between the floating patches of algae. She lands again, out of sight, between a couple of boulders. I have to wait.

Through my binoculars I spot her ten minutes later, walking up the gentle slope without the fish. It is peat country here, boggy grass and heather on metres-thick layers of peat, far away from any roads or houses, with sheep the only representatives of civilization. Uphill and more than 100 metres from the shore now, she halts, looks around her, and disappears into an inconspicuous crack in the peat. She is gone, nothing moves any more, it is the end of the story, and cramp in my leg is calling for a quiet exit and some lunch.

Female otter carrying a cub to the sea.

Next day, carefully watching from a distance, I see Wiebka charging down that same slope again. She is heading for the water – but this time she carries a small, fluffy bundle in her mouth, my binoculars showing it clearly. It is a small otter cub, black, baby-faced, but large enough so that the mother has to carry her head high, looking quite unbalanced. I am quite sure that this little one has never experienced water before, but nevertheless, Wiebka dives straight in with her cub and under, for at least five seconds that appear an eternity. Up on the surface again and still carrying the young one, she heads across the water to a tiny rocky island, a skerry with some clumps of grass, about 100 metres offshore. There the two disappear, safely in cover.

Brief moments later Wiebka heads back again to the shore across the waves, without her load, and she has left the cub on the little island. I keep at a distance and well out of sight, watching through my binoculars, safely downwind of her. Fast, fast she swims across, lands and gallops up the slope again, back into the same small hole in the peat. Almost immediately she reappears again, carrying a second cub, and I see the whole process repeat itself. The cub gets a dunking, is ferried across the surface, and left with the other one on the little skerry, out of my sight. And again Wiebka returns across the water, lands, gallops uphill to the hole in the peat, then disappears. But this time something different happens.

She takes quite a bit longer before re-emerging with a third cub. She walks a bit more slowly, down to the edge of the water, head up with the black bundle in her mouth, and there she stops between the rocks. She hesitates, drops the cub on the shore, hesitates, then dives in again, but without her baby. Off she swims to the little island, leaving the third cub behind close to the waterline, where I can see it through my field glasses.

Wiebka emerges from the sea on the little skerry, walks across the algae, between the rocks and grasses, and settles with the first two cubs that she had landed there. From my distance I catch just occasional glimpses. Her abandoned third cub begins to whistle. It is an un-mammal-like sound, like that of a bird, of a pipit, the desperate contact call between mother and offspring. If you know what the call means, it is heart rending, and if I can hear it in my far hiding place next to the rock, then Wiebka, who is closer to the cub and has better ears than I have, can hear it much louder still.

Half an hour passes, and the whistling calls of the abandoned cub are incessant. I see it crawling about near the water's edge between the algae – then the desperate wait becomes too much, and the cub launches itself in the water. Being fluffy, it almost floats on the surface, paddling along, whistling and whistling. This being almost certainly its first exposure to wetness, it must be terrified. Some 30 or 40 metres out it seems to lose its determination and it changes direction. Perhaps it loses sight of the little island where its mother and siblings are. Its fur seems to lose buoyancy; I see the little animal sinking lower in the water. Swimming in a circle, slower and slower, still whistling and whistling. Another five minutes of this, and its head disappears under the tiny waves. It goes down once, then up again. It disappears for good.

Of course I could have rescued it, somehow. But I don't. Years later I still often think about that decision, taken with a cold heart and leaving me feeling

Abandoned otter cub.

miserable with doubts. I rationalize that rescue does not provide a long-term solution, the cub is weeded out by nature, it would have had to go into a cage if I had picked it up. And yet ...

In my rationalizations I also wonder why Wiebka should allow her offspring to die, and why she inflicts this cruelty on a small cub. It is no unique occurrence either. Over the years, several other, similar observations of such apparent maternal neglect by an otter come my way. It is not really neglect, I think. It has all the appearance of regular otter behaviour, of deliberate abandonment.

The cold but most likely explanation lies in population regulation. Somehow, otter numbers have to be limited by the amount of food an animal can catch. A litter of young otters depends on the mother for food for a full year or even more, and she must work very hard to keep up the supply at all times. If she cannot catch enough fish, one or more of the cubs will have to die. I think

Otter cubs following their foraging mother.

that it would be difficult for her to choose on the spot which one that will be, especially as the cubs of a litter always seem to be close together. But there is this window of opportunity for her – this brief time when she is transferring cubs from one site to another, when she has the cubs well apart. It makes a desperately sad spectacle when a cub is abandoned, so the mother can manage to bring up the others to maturity, without having to kill one.

Perhaps it is rational and efficient, though very hard to watch without emotion. Incidentally, some other carnivores, such as spotted hyenas, when faced with a similar problem solve it through the cubs murdering each other. Among badgers some adults may kill cubs of closely related clan members. For lions, part of the litter just starves to death when conditions are bad.

I learn from my time with the Shetland otters that much of their behaviour and their vulnerabilities can be understood by their need for easy access to prey. The cubs need to be fed by a mother who can quickly bring back the fish, who does not lose too much energy while doing so, who is well insulated against a hostile environment, and who has access to easy fishing places in the habitat.

I become immersed in the otters' lives, I know them personally. Yet, I need a more satisfying explanation for their limitations. Obviously, what is required is knowledge about food, about fish. And alas, Shetland is wonderful for otters, but the endless sea makes insight into fish numbers and prey availability a far-off dream.

Along the Shetland coast I walk back to the bothy where I am staying. The boggy slope here has a large fresh peat face where Jimmy, the local postman, digs his peat for the winter months. Sheep move out of my way, a few gannets circle above the waves, rocketing down after a fish, then moving on to another site. A raft of eider ducks is moaning and cooing some distance offshore, the odd large black-backed gull calls overhead, eyeing me for my intentions. A skua whizzes past, on the lookout for victims that can be robbed. This world has everything, I have otters on my mind, I have memories. I still feel left with the question as to why there aren't larger numbers of otters; there is an answer here somewhere.

Food for the inexperienced: shore crab, rather unprofitable.

Otters. (Diana Brown)

CHAPTER 20

Otters in and around the garden

Otters on iced-over Loch Davan.

A LATE, PITCH-BLACK NIGHT in November, wind in the trees. I am driving slowly on a small road in north-east Scotland, near the River Dee. There is an antenna on the roof of my car, and soft bleeps in my earphones are the signals from a tiny radio transmitter, inside an otter, which is somewhere out there in the dark. Several weeks earlier, the transmitter had been carefully inserted in a small operation by a vet, after I had captured the otter in a large wooden box-trap. The antenna tells me in which direction it is, and I can hear from changes in the signal that the animal is moving. The otter is following a small tributary.

My drowsiness from lack of sleep is disturbed by a vehicle coming up to me from behind, which is unusual at this time of night here in rural Aberdeenshire, and annoying. Suddenly my tail switches on a blue light, and a 'Police' sign. The officer is polite, and asks what I am doing. Fair enough, of course, I know that there are salmon poachers around here at night. The policeman and I have a pleasant chat, I explain about the otter work, he tells me about the poachers. 'Why don't you tell us beforehand when you are out?' he asks. 'It saves us having to bother you.' Quite, but how on earth can I know where that otter is going to lead me, the animal having a range of some 40 or 80 kilometres of streams and river, anywhere in the area?

The episode reminds me of a recent incident of one of my students, who was similarly radio tracking an otter at about two in the morning in bright moonlight, not far from here along the River Don. Leon Durbin was wearing a balaclava in the dreadful cold, and stepped out of his car to pinpoint the radio signal of the otter with his hand aerial. Rather close to a remote house, the elderly lady living there looked out of her window to see, in the moonlight, a dark-clad potential terrorist wearing a balaclava, waving something that looked like a gun. Soon, four police cars arrived which pinned down Leon's vehicle. Leon was arrested and taken away, and only a nocturnal telephone call to his mother, down in England, reassured the police that their man was harmless.

My own house is well within the home range of the otter I am following. He is a large male, nicknamed Nelson. I have been on his tail for several weeks already. Tonight he is about ten kilometres from my house as the crow flies, and the stream he is following is an outflow from one of the lochs in the centre of my study area, into the River Dee. It is difficult to get even a glimpse of him in the dark, but I can predict that the otter is going to be crossing underneath the one small tarmac road here. I get ahead of Nelson and wait on the bridge where, peering into the night, I can just see a patch of water beneath.

A good ten minutes later the otter crosses underneath me, radio bleeps screaming in my earphones because of his close proximity, but I only catch a glimpse of the animal, and some ripples in the patch of open water. The radio bleeps subside again and, boringly, that is as much visual observation of an otter as I am getting that night. I guess that he will be heading towards the two quite large Dinnet lochs here, Kinord and Davan, where big reed beds are an attraction for any otter.

Interesting though, I also know now that Nelson will not be staying there. Like other otter males around here, he only uses the lochs for occasional nocturnal visits. Nelson really lives along the River Dee. For my study I sometimes have to catch otters in my large, wooden box-traps so I can provide them with a radio transmitter, and if I put a trap like that somewhere along the big river or near the bottom end of its tributaries, in the large majority of cases when I catch an otter it will be a male. The rivers are male habitat. In contrast, the lochs, reed beds and small streams are for the females, with or without their cubs.

The males must know their riverine habitat in astonishing detail, as I find on another occasion when I am trapping otters. On my daily check of traps, a beautiful, really large male otter is in the wooden box-trap under an old bridge, along the Beltie, another of the small tributaries to the River Dee. When I take the animal out of his confinement, both his eyes appear to be white, completely opaque and clouded over – he is totally blind. But from his excellent condition, his alertness and his thick fur just perfect, it is clear that he is thriving in his riverine habitat. How he manages it is an enigma. He is in his prime, but absolutely without any eyesight and he has managed this without falling foul

Loch Davan reed beds, prime otter habitat.

of traffic, people or dogs. It can only mean that he knows his area extremely well, with the sounds, 'texture' and smells of rivers and streams in every detail.

In the early morning after my nocturnal encounter with the local police force I am at my usual observation place along the shore of Loch Davan, hoping to meet up again with Nelson, as he was heading in that direction. A vain hope though, as my radio receiver tells me he is nowhere near. Leaving him to his devices, I settle down in the early sunshine, and slowly scan the surface of the loch with my binoculars, enjoying the various species of ducks feeding there, a couple of mute swans, a few greylag geese. The big hill, Morven, towers above the yellow reeds and birches. In summer I would have expected an osprey fishing here, but during these wintery chilly days it is plunging the waters of warmer climes in Africa. Just now my attention goes to a quiet part of the loch that ducks are avoiding suspiciously, and within a couple of minutes I have an otter there in my field of view, just its head, leisurely peering around.

Seconds later it disappears again with the characteristic tail flip, gone from the surface. Its head looked smallish and rather slim, and the animal shows none of the behaviours that, to me, identifies a male, such as its tail floating behind, head slightly out of the water, conspicuous, and swimming as if heading for some distant destination. No, my present animal is spot-diving, and I am quite confident it is a female, just as I expected here.

Strange: why should the two sexes keep out of each other's way, in different habitats? Is it anything to do with females keeping young cubs away from the males, who are known to attack? Or do males stick to the rivers where they are likely to encounter larger salmon, leaving the smaller fish, such as the eels of the lochs, to the somewhat smaller females?

Eels in Scotland, a favourite prey for otters.

Up she comes again, the female I am watching. She floats for some ten seconds, then dives, more or less in the same spot. A dozen more such dives and she emerges with an eel wriggling around her mouth. This time she makes a bee-line for the nearest reed bed without any hesitation, and disappears, eel and all. She may have cubs there.

I am admiring the large reed beds from where I sit, contrasting their beautiful yellow lines against the dark brown of Morven Hill, and the grey-blue lake water in front. The reed beds are excellent cover for otters, and whenever I walk through them myself, I disappear into a world of my own. I can follow the narrow trails of otters, the deep holes where otters dive underneath the floating reed bed, the couches where otters have dragged stalks to make nests, where they curl up and rest.

An otter couch is no more than a kind of sleeping platform, quite open. But there are times when otters construct something much more elaborate, and they only do that in reed beds. On one occasion, when following the radio signal of one of our otter females through the reeds with a colleague, Paul Taylor, we decide that, for once and against our established routine, we should walk close to her to see what she is up to, because the signal has not moved for several days from one place in the thick of the vegetation. Fluctuations in the signal indicate that she is moving about. The location is quite a long distance from the open waters. Carefully, and with as little noise as possible, we get close to the large clump of reed stalks, the radio signal at full strength and clearly coming from that one clump. She is in a 'nest', a rectangular ball of about a metre high, made entirely of reeds, almost as a bird would do it, such as a wren or a magpie. We can see a side entrance, and twice I hear the squeak of a cub.

This is the place where she gave birth, a 'natal holt', and when I see her again several weeks later she has two little ones with her. A few metres distance from the holt in the reeds is a large pile of spraints.

Getting my thoughts back again to where I am now, watching the reed bed from across the water, with ducks in front, it is obvious that the ducks are avoiding the otter. My mind wanders again, this time to a previous observation, when I saw an otter dive some 20 metres from a mallard, only to come up underneath the duck at exactly the time that the bird exploded out of the water. Lucky duck, as there was no doubt what was in the otter's mind. So I am always alert to the possibility of a bit of duck sport for the otter, and it is not surprising that the ducks swimming about in the loch between me and the reed bed keep well clear of the predator. No luck for the otter this time.

Less than a year ago, in midwinter snow, I followed the tracks of an otter as it was moving away from the pond in my garden, dragging one of my own beloved white ducks for more than half a kilometre before eating it. It could possibly have been the very same otter I am watching here today, because the home ranges of otters here at the loch also comfortably cover my garden. A little stream runs through the garden to the river down below, a watery connection between the pond and the rest of the otters' range. These animals cover large distances of stream and riverbank, females regularly 40 kilometres or more, males twice as much, all that way even within a couple of days.

One thing that my radio tracking of otters tells me is that not only do they use very long stretches of river or stream, but they also spend a lot of time in very unlikely-looking tiny trickles of streams. These are called 'burns' here in Scotland, where often otters find small fish such as sticklebacks and small trout, and frogs. Even more surprisingly, otters cover many kilometres overland, between watersheds or to ponds, getting themselves far away from streams. I find tracks of otters here in Scotland in places where no one would expect them. Sometimes there are otter tracks in places where the legendary 'big black cat' was seen just before. I know what I believe of those much-published cat sightings, of those 'observations' by people at night, often at times just after the pub closed. Otters.

Right at the moment, my animal is somewhere in the tangle of the large reed bed I am watching, chewing up the eel. Usually the predator starts with the head, but often before that it thoroughly rubs the eel on the ground or plants to take off some of the slime. I am always pleased to find such places with eel-slime along the edge of the water, as another small window into the lives of these animals. Eels in these waters are probably the most important prey for otters, for the simple reason that, oily as they are, they provide so many calories. And to an otter, those calories are hugely important. Not only that, but eels are also very easy to catch, as I notice myself when scuba-diving in Scottish rivers and small lochs. When eels are resting, part-covered in the muddy bottom, one can often easily touch them by hand.

I realized the importance of calories to otters because some years ago, with my colleagues, I had studied otter diving in a large tank next to the institute where I worked. I had always been suspicious of ideas about otters' ability to protect themselves against the cold when diving – they have marvellous fur but, unlike seals, otters have no blubber underneath that, and there just cannot be that much air captured inside the fur when underwater. Insulation is therefore limited, so the animals must get pretty cold when swimming and diving for any length of time. To make up for that, they just need to eat masses of calories.

At the institute I had been able to arrange for the building of an enormous tank, where we could measure the amount of oxygen otters used from a large air-bell above the surface when they came up after diving. Their oxygen consumption turned out to be more than expected, and it was dependent mostly on the temperature of the water. The colder it was, the more oxygen the animals needed. Using more oxygen means that they are using up more calories from their food. Needing a lot of calories means that otters have to be very efficient when fishing, by capturing a lot of fish in a short time. Otherwise, the fishing effort loses more calories than it gains. It explains their need for eating fatty prey such as eels, and for catching prey without having to dive for long.

The animal I am watching, here in Loch Davan in front of me, needs to capture an average-sized eel well within 20 minutes if it wants to break even on calories spent and gained. If it doesn't catch fish quickly enough, it will lose calories and suffer, and certainly there will be no energy for reproduction – she has to fish elsewhere to get enough for that.

Recently, I notice here in the loch that otters are now spending longer in the water for every eel they catch. They take longer than when I first started watching them. Eels have been fewer in recent years, and other fish, such as perch, are not as rich as eels, so otters mostly ignore them. It is not surprising that I haven't seen otter cubs born here along the lochs in the last few years.

Male otter diving.

Otter vibrissae, essential equipment during fishing.

In my earlier days there were litters every year, at least one, sometimes two or even three. The dearth of eels, possibly caused by climate change, is taking its toll.

While I am following my otters, a colleague, David Carss, keeps a close eye on fish numbers in these same lochs and rivers, on trout and eels and others, and how much these fish populations produce. He uses electro-fishing, catching all fish in small sites using minor electric currents. Between us, we obtain pretty strong evidence that otters take a quite large proportion of the available fish. It is not surprising that they may be pushed for food.

Wherever I turn with these animals, the explanations for otter numbers and their social life, their populations and many other behaviour features, they always seem to relate to food. I admit to being blinkered by this to some extent, as of course I realize that other threats in their environment also play a role. I am thinking about this when, one night on foot, I am following an otter with its radio transmitter in the pitch dark in Scotland and it decides to go walkabout on one of the roads through the study area. Grumbling to myself about the madness of these animals, I watch the otter as it carefully follows the white central line of the tarmac road, for hundreds of metres into the darkest of nights, with me watching and dreading the possibility of vehicle traffic. Fortunately the otter abandons its foolishness in time, but the incident clearly illustrated the dangers of road traffic to the species, and large numbers of them get killed every year. What this does to their populations, no one knows.

And otters face other dangers, lurking in the dark and kept out of sight of conservationists. Here, on the very same Scottish estate where I am watching otters at the lochs, it is not many years ago that I found the animals' holt targeted by gamekeepers using strychnine. Elsewhere in the area I find dead otters full of shot, one killed by a 0.22 rifle, and in my first years here I even come across several gin traps. Even in Britain, man was and is a predatory force to be reckoned with. Yet, otters are doing quite well. They are magnificent stars in the beautiful Scottish environment. And as to the answer to that old question of mine – 'Why aren't there any more otters here?' – I have to agree with the opinion of the local fishermen. 'There aren't enough fish in the waters.'

CHAPTER 21

Otters, crocodiles and orcas

Spotted-necked otter in Lake Victoria – camouflage!

AFTER SHETLAND, and after the years with the otters around our house and the rivers and lochs, I feel the need to see how other kinds of otter fare abroad. There are different species of otters almost anywhere in the world (except in Australia and Antarctica), all looking remarkably similar, and living in very similar kinds of place. Except that often in these other countries otters have to put up with much more prominent predators.

It is midnight under a small crescent moon, with a few cicadas singing away in the trees above the house where I am staying. Outside the net over my bed, mosquitoes are buzzing, thousands of them. I am awake, thinking about life in this hot and humid place, listening to the tropical sounds. The house is far from other habitation, close to the water's edge on Rubondo, an island in Lake Victoria, in the middle of Africa. Rubondo is covered in thick forest, with dense shrub, trees and even palms right up to the edge of the water. There are occasional rocky outcrops into the lake, creating small peninsulas. It all feels like the tropics as they should be, forested and wild.

From my bed I listen; the windows are open and covered in (useless) mosquito gauze. In the far distance on the mainland I can hear African drums, their sounds floating softly across the enormous lake. A hippo roars, maybe a kilometre away. Every so often I hear a loud bang, and it takes me some time to realize what it is: the tail of a large animal, hitting the lake waters with explosive force. I am hearing crocodiles, and large ones at that.

The sound reminds me of what happened earlier that day, when I am battling my way through the vegetation along the water's edge. Suddenly a really large croc comes charging at me from higher up on land, a beast about 2½ metres long which I have obviously disturbed, rushing past on its way to the lake, just a couple of metres in front of me.

Crocodiles scare me and make me feel highly vulnerable, especially of course when I am swimming in the lake. They are the enemy, as they are also for the otters I come to study here. But they may be an important force in the environment, and an answer to one of the many questions I have about these Lake Victoria otters. Gradually I drift off to sleep.

The heat arrives early in the morning, and rather suddenly. The lonely house where I stay, called Kageye, feels its age; it is dank and mouldy, having grown into its surroundings under the massive forest trees. I can see the lake from where I have my breakfast, and the house is surrounded by life, one of those places where I need to have my binoculars ready at all times. Chameleons climb the branches near the window. Still somewhat drowsy, I fill a Thermos with cold water – causing the bedraggled face of a dormouse to pop up from the flask. It should be thankful that I am not taking hot coffee with me.

Following the shore brings me to one of the rocky promontories, where I expect otters – and I find them too, so I can sit down carefully, and watch. The species here is the spotted-necked otter, a bit smaller than the Eurasian one and rather slimmer, more mobile, agile and faster. I can see seven of them together, with the animals walking about on the rocks, using their sprainting site on the flat rocks, and going for brief swims and dives. They are common along this shore, and it is fascinating to see this different species and to compare its behaviour to the one I am used to at home.

Suddenly, in one go, the spotted-necks are off, all seven of them, away along the densely wooded edge of the bay. What strikes me immediately is that the otters each fish on their own, apart from each other scattered along the rocky and densely shrubby shore, in and out of the branches. They stay close to land, often emerging with a small fish, then disappear again in the vegetation. There is no fishing together, no obvious contact between them. But once they decide to push off to a point somewhere further along the coast I see a clear change. Heading for the next promontory across the bay, they bunch tightly together and look almost like a single animal, swimming fast across the open water quite far out, leaving me to contemplate my misery in the dense shrub and trees, with no otter in sight any more.

Clouds of mosquitoes dance over the water, reminding me that it is almost a new moon when I have to breathe through my handkerchief to escape breathing in the hordes of these insects. Fortunately they are only chironomids, not biting mosquitoes. Regularly once per month, chironomid numbers erupt here catastrophically. I get down to the rocks of the promontory and collect a batch of the otters' fresh spraints, so I can sort out later what they have been eating.

Clearly, these otters are gregarious. They are a very social group, and a tight bunch is indeed the last thing I see of them this morning, swimming close together, disappearing into the distance where I can only follow slowly. I am reminded of the noises I heard last night from the waves in that direction: the loud slapping of crocodile tails. Could their grouping behaviour, their bunching together, their activity in daylight, have anything to do with the crocodiles? In daytime, the crocs are mostly basking ashore. Interestingly, I hear from colleagues that this same otter species, the spotted-necked, is just as solitary and nocturnal in streams and rivers in South Africa as my Eurasian otters are in inland Britain. That is in the absence of crocodiles. Here in Lake Victoria they group, and fish in daytime.

I return to my Rubondo fieldwork again. Walking, sometimes creeping along the densely wooded shore of the lake in the direction of where my otters disappeared, I arrive at a small, sandy beach. A herd of hippo is basking in the morning sun, a couple of dozen of them, with two egrets walking in between. There is no sign of otters, but the hippo are fun to watch from behind a large bush. I enjoy it until such time as I decide to continue my walk to where I think the otters might have gone. My getting up causes a gigantic splash of the entire herd of enormous hippo bodies into the lake. I would not want to go for a swim here, just now – hippo have to be treated with considerable respect.

Struggling on through the thick bush and tall forests, I pass through a rather damp, open area with a lot of long grass, elephant grass, which is much taller than I am. It is difficult to see ahead of me. Watching my step while remembering the large crocodile that I disturbed yesterday, I get a shock when an animal crashes away from almost under my feet. It stops briefly, and I just manage to get a good glimpse of a large otter before it disappears totally. It looks quite different from the spotted-necked otters along the lake-shore: this one is much larger with a broad head, and white underneath. Next to the place where

Two spotted-necked otters on the Lake Victoria shore (one melanistic).

it had been curled up asleep there is a nice collection of scats, immediately showing themselves to be full of bits of freshwater crab. I have just seen a Cape clawless otter, the other species here. Throughout Africa, crabs are this otter's speciality. I am so delighted I am whistling to myself: rarely does so much fun come my way all at the same time.

Moving about is difficult, at least on land. Rangers from the national park who are in charge of Rubondo Island have canoes, and next day they offer to take me around, along the coast, otter watching. They also want to show me what they do, what they are up against. Four rangers are doing the paddling and, because I am in the prow, otter spotting is a doddle, with the rangers quiet and experts at keeping their distance. The spotted-necked otters we meet do notice us every so often, but they seem to be little bothered by us. I get beautiful views of the otters' funny little, square-ish heads with long, large-spotted necks poking out of the water. Small as this species is, their faces still remind me more of the South American giant otter than of the (to me) most familiar Eurasian species. Like those South Americans, they are so much more gregarious – at least they are here, in Lake Victoria. I am getting some lovely data on their feeding.

My thoughts in the canoe are suddenly disturbed by a hushed shout in Swahili from one of the rangers, '*Watu!*' (people). The boat steers into the wooded bank, rangers jump out and run, and in no time they catch six fishing people. They are poachers, because Rubondo island is a national park. The six lads, in rags, have their hands tied and are put in their own canoe, which was hidden under the bushes. While the rangers search for and collect illegal fishing nets, I am in charge of the prisoners; I am armed with a *panga* (machete), but I have no idea what I should do if one of the men should try to escape. Fortunately, they do not seem to mind their predicament at all – they are chatting, they just accept the fact that they'll go to prison for a bit, where they'll be comparatively comfortable and well fed. But they will lose their canoe and their nets. On our way back, the poachers have to do the paddling for the two canoes, their own and ours, and that is the end of my otter watching for today.

The fish that people catch in their gill-nets, stretched underwater between the trees and bushes, is mostly *Tilapia*, quite large and delicious to eat – introduced into the lake and not native here. The native fish fauna consists largely of several hundred species of tiny fish, mostly *Haplochromis*, all closely related cichlids, which probably are the otters' mainstay. Unfortunately, not so many years ago somebody in Uganda decided that these little things weren't good enough for people, and (literally) tipped a bucket full of Nile perch *Lates niloticus* into the gigantic Lake Victoria, where the new species multiplied beyond belief. Nile perch grow to an enormous size, they multiplied into millions, they are voracious and cause the extinction of the large majority of small cichlids. Now, a substantial fishing industry has developed in Lake Victoria based on the Nile perch. It is a rather greasy and not very tasty fish, but it does feed millions of people. It will not be doing much good to the spotted-necked otters, whose *Haplochromis* prey is being wiped out.

Another consequence of the introduction of Nile perch is at the moment plaguing me here daily. It is almost new moon, when the small midges, chironomids, swarm for a few days. They black out the sky, making breathing

Chironomid swarm on Lake Victoria.

difficult for anyone out and about or even inside their house, covering car windscreens in a layer of a kind of midge porridge, and endangering aircraft. These little things, six different species of them, used to be eaten by the millions of *Haplochromis* fish that were in the lake, but no longer, as the Nile perch has eaten the small fish, and it is not interested in the chironomids itself. Lake Victoria puts up a giant display of ecological disaster as caused by people.

That night I sit on my little private beach near the house on Rubondo island, surrounded by glow-worms in the dense shrubs, looking at the stars. It is as close to paradise as I will ever be, with the smells of the forest, the sounds of insects, and looking at the many lights of small fishing boats in the far distance across the lake. Again I hear crocodiles slap their tails on the surface, a reminder that, for the spotted-necked otters, Lake Victoria may not be such a paradise after all. Perhaps their group-living protects them from the evils lurking in the waves: many eyes are better at spotting danger than merely two.

Years later, and many worlds away from Rubondo. I am again watching another kind of otter. My view over the waters where I am now is spectacularly different, beautiful and totally other. The otter species, in front of where I am hiding just now, is only distantly related to the animals I saw in the huge African Lake Victoria. But as in Africa, here too there are two different species, here too is the threat of large predators looming over them.

Alaska is grand in every way, and the Pacific coast along the Prince William Sound is awe-inspiring. My mind tries to dismiss the picture-postcard image effects of the large, snow-capped mountains and glaciers as backdrop for blue waters. Such images may be cheap and rather chocolate-box-like, yet I feel overwhelmed. The mountains drop into the sea with steep, wet forests of hemlock and firs, and I am very much aware of the mass of wildlife inside them.

The North American river otter that I am now watching dives in the waves of the Pacific Ocean while I stay quiet behind a large log, only about 20 metres away. A brightly coloured bald eagle sits high on a branch above the waters, but I keep my binoculars firmly focused on the ripples in front of me. A soundless dive, a tail flip and just a few bubbles – then, some 23 seconds after the tail disappeared under the surface, the otter pops up again like a cork. Just its head shows while it is scanning the scenery, chewing on a quite small, obviously bottom-living fish of a species unbeknownst to me. I can hear the smacking noises, and if I weren't alert to the different shape of this North American river otter's nose, I would have felt quite like being back again in Shetland with the Eurasian otter, despite the totally different scenery.

Just an hour or so earlier, I was dropped off here by a small open metal boat, a Boston whaler, driven by two lads who are helping our expedition. I am left to potter around along the shores and, while they go fishing for snappers for supper, I am otter watching. It is wild, wild country here, with no other people around for many miles; I am aware of a glacier and mountains above me, it is cold and overpowering.

The river otter, here a marine animal, drifts away again. It follows the shore and leaves me obliged to follow on land. Along a small beach I notice the large tracks of a bear, perhaps the same black one I had seen earlier on a slope above the forest. A black-tailed deer scurries off while I concentrate on not disturbing my quarry. The bald eagle watches, and almost casually I notice that another otter here, a different species – the sea otter – is floating past a bit further out on the water. Sea otters, incidentally, are everywhere along this coast. They are mostly on their own although they are the most gregarious of all otters, sometimes congregating in 'rafts' of scores of animals. They are abundant, despite the presence of what is probably their main threat in these beautiful waters, orcas, or killer whales.

After a few hours, my otter watching is disturbed by the sound of the returning Boston whaler. The boys are back with a good catch for our supper,

The coast of Prince William Sound, Alaska.

Sea otter in Prince William Sound.

and I join them aiming for the warmth of our temporary base, the yacht *Babkin*. The yacht is a good hour away, so we are moving fast through the straits and narrows and open waters of this wide Alaskan seascape. On the way we spot two orcas quite close, in the distance a humpback whale is breaching, and I keep making notes of lots of sea birds. What more could one want out of life?

Somewhere along our course, in a good place well offshore, the boys suggest we put out a line for halibut, before it gets too dark. I am handed the tackle, shown how to bait it, and down it all goes, really deep. With my extremely limited fishing experience I expect very little, but within minutes of the bait hitting bottom there is a tug on the line. After what seems ages of very hard work and with a lot of help, I heave an enormous halibut alongside, all 48 kilograms of it. It must be the proudest moment of my entire (admittedly extremely modest) fishing career.

On the *Babkin* the crew and the girls, including my student Gail Blundell, have been processing several otters, caught in traps set by our team. We, that is the team, are here in the Prince William Sound on the yacht to catch otters, measure them and take samples, and provide them with internal radio transmitters. Ten years earlier, a horrendous oil spill blotted out life along these shores. The company responsible, Exxon, is financing our research as well as that of others to establish whether the effects of the oil spill are still noticeable after all this time. Vast amounts of money are being spent, hence our grandiose yacht, flights to and fro and radio tracking with seaplanes, a unique opportunity for wildlife research. I am still somewhat nonplussed about it all, as just now it seems impossible to imagine this huge and unbelievably beautiful country under a layer of oil – but it did happen.

Later that evening I lie in my bunk, listening to the waves lapping against the sides and thinking about my simple observations, of river otters feeding, of sea otters passing by. For me, such times provide further building blocks to my small understanding of competition between species, of differences in sociality, and of evolution. Small insights, but they make deep impressions on me against the enormous background tapestry of Alaska.

Next morning I sit on deck while we are still anchored in a magnificent bay. On the slope above the forests I watch a black bear moving about, a bit too far to see exactly what it is up to. Miles away, an avalanche produces a deep, rather disquieting rumble. A belted kingfisher flies past, and a few guillemots bob up and down nearby, a black one and a pigeon guillemot. Today I am accompanying the girls on some further otter trapping for their research projects.

After breakfast we get into the Boston whaler, and dart across the waters to where a radio signal tells us that an otter has been caught in one of the traps. I am very impressed with the capture procedure, the darting and getting the otter out of the trap with a minimum of distress, sampling its blood, and operating and inserting a radio transmitter. It is expert work, done extremely professionally by Gail and her colleagues – she is a former veterinarian. What really excites me, though, is the place where all this is done.

It all happens on land beside a gigantic otter den, a holt, inside the steep, wet, densely wooded slope. It is one of many dens here, and I never saw anything as large and extensive as the holts of this American otter species here. It is a collection of very large and wide pits from which tunnels branch out, some so deep that I can stand in them; there are water pools and small streams coming through, and large heaps of scats everywhere. Otter trails run down the steep slopes to the sea, also from one entrance to another. Obviously this holt has been used for many years, by large numbers of river otters.

I realize that with this huge den I am looking at a spectacular demonstration of otter gregarious living, and while I am clambering around, in and over it, I think of the observations people tell me about, of a dozen or more males going round jointly in 'bachelor packs'. Females are quite solitary when they are fishing, but overall the American river otter here is much more social than the one I know from Britain. And in addition, across the waves below the slope there are the sea otters, the other species, also much more inclined to do things together, sometimes seen in large 'rafts' of scores of animals. These animals are gregarious when out in the open waters, and in the forest around me, areas where they are vulnerable. Predators, such as the orcas, bears, wolves and lynx here, are no threat to the Eurasian species at home. Here, predators are a large presence.

The scientists working next to me on the otter, on a makeshift operating platform, use the latest modern equipment. They are well provided for: they have many radio transmitters and recorders, the latest computerized backup, dissecting equipment, and highly specialized expertise of tranquillizers and operation procedures. And of course they have the entire set-up of the yacht, being provisioned by air with small seaplanes. As a guest I feel like a country bumpkin, with my binoculars and the occasional radio transmitter at home as my most advanced piece of equipment, watching my otters in their habitat without the interference of me as observer.

Here in Alaska I cannot help but quietly make comparisons with my unassuming fieldwork on Rubondo in Lake Victoria, with the canoeing rangers. I see here all the trimmings of American advanced technology and specialist science. I am not envious, because I also feel that this new direction of science is missing out on something. People don't often watch animal behaviour in the field any more, they let equipment, laboratory and computing take over, and they develop techniques to record data digitally and automatically. Yet for me, just observing is still important. I get my thrills from watching, from getting ideas and hypotheses by looking through my binoculars. It needs all sorts to make a world.

Years later, it is 2005. Lots of caiman, capybaras and masses of birds see Jane and me floating along the wide Rio Negro in Brazil's Pantanal, in a large canoe. In a hot, tropical abundance of life we are a world away from the otters at Loch Davan and Shetland. The river drifts slowly along the banks and their vegetation, and I am thrilled to see the huge animal diversity. I am told that there are some 260 species of fish in this river. Of course, what I am really after, in my own blinkered way, are the otters.

Again, there are two species here, not only the spectacular giant otter but also the neotropical otter which looks very similar to my animals at home, the Eurasian otter. It is almost weird that the animal I am watching from the canoe in this other world looks and often behaves so similarly to the otter in the Scottish lochs. Solitary, the neotropical otter dives quietly underneath floating masses of water hyacinths, often catching bottom-living fish, even quite large catfish. It is active in daytime as are the Eurasian otters around the coast in Shetland, although in streams and lochs the animals are mostly nocturnal. Otherwise, the neotropical otter fishes, rests and scent-marks in a way largely identical to what I am used to at home.

I think about how fascinating it would be to study this animal in detail, because I also notice some tantalizing differences. For one thing, as in Africa and in Alaska, there are more predators here, not only the caiman (the local crocodile), the large cats (even jaguars) and eagles, but also, and perhaps more importantly, small enemies: piranhas, highly dangerous, quite small but sharp-toothed fish that may attack en masse. Many of the neotropical otters I see in the Rio Negro have scars; some have large ones, and often part of their tail is bitten off. These are a sure sign of piranhas, the locals tell me.

Also, there are the much larger other ones, the spectacular giant otters, gregarious, boisterous and noisy, often in the same waters, and I see the two

Cayman in the Pantanal, a danger to an otter.

Neotropical otter with catfish.

species studiously ignoring each other. With all these other environmental challenges to the neotropical otter, and with the abundance of different fish species to feed on, in much warmer waters, it seems unlikely that they are simply food-limited, as I think my otters at home to be. There is a great deal to be found out here.

My most important concern with these two otters in the Pantanal would bring me straight back to one of the original forces behind my entire research career. It lies in the contrast between the two otter species. The question of why the neotropical ones are solitary, and the giant otters gregarious, is a problem that might shed light on the evolution of societies.

Perhaps, and here I am guessing, the reason for the gregariousness of the giant otter is similar to that of the Eurasian badgers, which also often live in groups. The badgers' food is found in distinct food-rich patches, which are available to the animals at different times. To survive, a badger needs to defend an area with several food patches. But within the boundaries of such an area, a number of individuals can live without competing, because of that patchiness. And once several, or even many, badgers live jointly in a territory, co-operation between them would be a clear advantage. Giant otters are known to co-operate with group members in many ways.

If giant and neotropical otters feed on different prey, perhaps using deep and shallow rivers, oxbows and (during the rains) flooded forests to different extents, their different social lives could be understood in the same way as we understand badgers, through the distribution of their resources. But we will only know after a proper study, perhaps here in the Pantanal. In the meantime, I still have many other problems to tackle in the Eurasian species.

CHAPTER 22

Three monks, calls of gibbons, and otters in Thailand

Huay Kha Kheng tributary.

Walking on the sandy flats of a large and mostly dry riverbed through the hills, I am surrounded by huge, tall rainforest. It is early in the day before the heat comes down, and I feel and hear the forest waking up, despite an early-morning mist emphasizing a deep silence. The dark wall of trees comes right down to the riverbed. It minimalizes me.

I am miles away from any sign of civilization, in a gigantic nature reserve along the River Huay Kha Kaeng (the 'River Kwai'), in Thailand on the border with Myanmar. The country feels, and is, enormous. Lifting my eyes every so often into the mists from the shallow waters and the sand where I am looking for animal tracks, I expect an elephant, a tiger or at least some sambar deer ahead of me. Instead, only half an hour into my walk some strange and only slightly moving shapes detach themselves from the mist, in a small, dense group on the sand. People.

Three monks are sitting on the sand, close to one of the many dry-season pools, talking quietly and breakfasting from their food bowls. In their thin, yellow-brown robes, with shaven heads, they can have little comfort in the early cold. It is a moving scene in the endless rainforest. I pass them, we greet with a *wai*, and we each disappear in the mist.

Slowly, sun begins to catch some of the treetops, and a world stirs. Green pea fowl roost high up above the mist, and several start to call, with a loud reverberation from their fabulous colours. The birds appear to be made for this primeval forest. I walk on, trying to focus on the sands, to spot animal tracks from last night's activities here, but it is very difficult to concentrate. It is not danger that distracts me, although there are large cats and elephants about. It is the sheer multitude of wildlife that makes my life complicated.

One sound more than any other gives me the shivers, a sound with very deep emotion, like some of Mahler's music. Up in one of the highest trees is an ape, a white-handed gibbon. He calls, very melodically singing in high, endlessly varied notes that carry vast distances. Several replies come from all directions. I can only compare the concert with the howling of wolves, although the gibbon is much more melodious. The broad riverbed winding through the impenetrable forests, the last morning mists, and then this sound, it is totally overwhelming and almost too much to take in. There is a mood of vastness, a world of unfamiliar sounds without echoes.

I am here to study otters, but it is difficult to stay blinkered, faced with such a rich tapestry of environment. Of course there are otter tracks in many places here, but how can I stay concentrated on my subject when there are fish eagles overhead, when troops of macaques claim the entire riverbed, when masses of birds demand identification and the forest makes me feel so insignificant? There are tracks of different species of civet cats, of a leopard, of mongooses and monkeys. It is fabulous.

Along this river and its small tributaries there are three different species of otter, which is what brings me here. There is the Eurasian otter, the one I know at home, but also troupes of small-clawed otter, the one that is often kept in zoos, and the larger smooth-coated otter (which is its rather stuffy official name – I always call it the smooth otter).

Reflecting on some of my previous research, I am trying to sort out how these three can exist together in the same area. Don't they compete with each

Monks along the river.

other, for instance over food? Or is competition not an issue, along this river where someone found over 100 different species of fish, as well as more than 40 different kinds of amphibian, as well as other otter delicacies? And are all three otters equally solitary, or gregarious?

Some of the Thai people who help me here are excellent naturalists, and they know the area well. Out in the bush, however, I find them rather noisy, so I like to walk and watch on my own, discover the tracks, and quietly watch whatever comes my way. My Thai friends are also shy, reluctant to commit themselves to naming animals or tracks that we see, although they know much, much more about them than I do. I am often aware that when I come across the track of some new species, they are reluctant to name it. I may suggest it is a civet or whatever, they will nod wisely and say 'Yes, yes, a civet, you are right', despite knowing full well that it is a quite different animal.

The otter tracks in the sand are often beautifully clear, and it is not too difficult to distinguish species. It seems ridiculously small-minded of me here to concentrate on just otters in these huge surroundings, where there are at least another 20 different kinds of carnivore, as well as deer, wild cattle, elephants, masses of birds and whatever. However, I have to be blinkered, otherwise I would not get anywhere with my science.

Along a muddy pool I find the familiar spoor of my old friend the Eurasian otter, in and out of the water, as is its wont in the Scottish countryside. Somewhat further along, an enormous rock on the bank draws my eye, and I notice small game paths in the tall grass nearby. Following one, it leads me to

Tracks of three otter species in the Huay Kha Kheng sands.

a flat area on top of the rock with many large otter scats, full of crab remains: probably the work of small-clawed otters. Keeping an eye out for snakes I collect the spraints while studying the otter footprints. All these small details provide data.

Later that day I sit in front of my bamboo hut high above a bend in the river, one of the huts we are staying in. My Thai friends and I are enjoying the view with a beer, looking out over the forest and the river, pointing out the many birds in the branches and a deer in the distance, talking about plans. Suddenly, the leader of our expedition, Sawai, quietly points at the riverbed and the large pool below the overgrown bank.

Three otters are busy there, and my binoculars tell me that they are the largest species, the smooth otter. They are foraging, and from our grandstand viewpoint we can see every detail. At first, the animals appear to act independently, but we soon see that something special is happening. Swimming and diving about two metres apart in synchrony, two of them cross the pool again and again, often emerging at the other end with a fish. Together, they are driving shoals of fish into the shallows, in a neat show of co-operation. One otter alone would have had much greater difficulty doing this.

It is very exciting. I have never seen such a wonderful performance before, and I see it as a building block for the evolutionary structure of social organization of these otters. It is an almost casual observation, but deeply significant. My Thai friends are less taken aback by this than I am, because something else is going on at the same time, also spectacular.

Near, above and next to the otters, three conspicuous birds are busy fluttering about: they are kingfishers and a heron, almost distracting from the excitement about the otters themselves. The most striking of these is a stork-billed kingfisher, a large and somewhat clumsy bird with a lot of orange and blue, and a huge bright red bill. There is also a common kingfisher, a fast and bright button of a bird, the same as the one in Europe, and a small, mostly brown Chinese pond-heron. They are diving and dashing after the fish that are chased or disturbed by the otters. The entire scene is one of colour and confusion. This also was an observation new to me, of birds using the otters for their hunt, a kind of commensalism just like the chanting goshawk and honey badger I saw in the Kalahari.

Next day I meet smooth otters again when I am following the sand flats in the riverbed – perhaps they are the same ones. Sitting down on the sand I watch them through binoculars from a distance, and again I am struck by how much more gregarious they are than the Eurasian otter; there is more joint foraging than solitary. I suddenly notice that one of them sits up on its haunches, looking round – something I never saw the Eurasians do in the wild at home, although one often sees it from Eurasian otters in captivity (and on postcard pictures). I must admit that other observers have occasionally seen Eurasian otters also sit up like this in nature, but obviously it is rare in that species, whereas I now know that smooth otters do it often.

The river through the tropical forests here in Thailand is a revelation about many sides of the otters' existence, with new behaviours that I see in my wanderings along the stream bed. The three species repeatedly rub it into me

Spraint site of a smooth otter, attracting one of the many Thai butterflies.

that they are so different. They eat mostly different prey (though also a lot of the same), they live in somewhat different parts of the enormous watershed, and they do things differently on their spraint sites. To my great satisfaction, some of the tracking skills I learned in the Kalahari come in useful.

I find the big, floppy footprints of the smooth otter mostly where the river landscapes are wide, large and open, and those of the smaller species near the more rapid flows, higher up in the watershed. Large, smelly and messy spraint sites are those of smooth otters, the drier sites somewhat away from the water and full of bits of crab are from the small-clawed otters. Small fishy spraints on rocks in or near the water have been produced by the Eurasian one. The tracks tell me their stories repeatedly.

In dense, long-grass vegetation I watch a group of small-clawed otters, I think there are six, but I cannot be sure because of their chaotic messing about. Under the debris from a previous flood there appear to be quite a few freshwater crabs hiding: the otters poke around and under them with their sensitive, almost clawless fingers. I see and hear them, noisily cracking open, munching and chewing the small crabs.

All this is against a tapestry of masses of magnificent trees, not far from the carcase of a sambar deer killed by a tiger along the riverbed, and from elephant tracks in the dry riverbed. A herd of elephant has walked along the sands, and in some muddy places their footprints show every detail, such as their nails and the small cracks in the soles of their feet. There are tiger footprints actually in the elephant prints, then a common otter track crosses both their paths. My cup floweth over, again.

One thing I learn about otters in the midst of this abundance is a need for humility with my suggestions and conclusions. Elsewhere, I thought that perhaps when otters live in groups it protects them, against crocodiles or against killer whales. And yes, it appears likely that some otters' group-living and co-operation not only aids against being preyed upon, but also with feeding, and perhaps they even help each other making dens. But here among the Thai otters, two of the species walk, swim, feed and sleep in groups, whereas one species does all this on its solitary own – and all three do this in the same area under the noses of the same many predators, of tigers, leopards and other cats, of people hunting. Otters are officially protected by law, but the threat of poachers with their traps is there. The gregarious species, the smooth otter and the small-clawed otter, and the more solitary Eurasian one, all appear to be doing similarly well under this. Again, I am made aware how much more there is to be understood.

All the different otter species across the world look rather similar – they all are very obviously 'otter'. With their appearance, their slender build, their lack of fat under the skin, they all face an aquatic habitat that requires all of them to feed very efficiently in order to acquire more calories from their prey than they lose in the cold water. Despite such similarities, differences between the three species here are large – differences in behaviour and ecology, between the habitats and foods they select. One concentrates on crabs, one eats small fish and amphibians, one is mostly after larger fishes. Hence environmental threats to their existence will differ for all these kinds of otter. They need to be studied much more closely, if conservation legislation and management is to be effective.

Living in groups is a pattern that has evolved in several quite different species of otter, in different genera, so it is likely to have evolved separately and independently. It is a fascinating adaptation to different demands from the environment, whether it be different kinds of food, predation, the construction of dens, or whatever. I see it here among the otters, just as I find it among the hyenas in Africa, or badgers in Europe.

CHAPTER 23

Down under: platypus, quolls and leeches

Platypus munching.

THERE IS A BLOB floating on the water in front of me, some ten metres away. From it, an endless series of concentric ripples is spreading out across the pool, shaded by enormous eucalyptus trees. There is action on the surface, with two small black eyes peering in my direction. I am very aware that if I move it will all disappear. The problem is that I am under considerable mental strain to move, prompted by an army of leeches galloping towards me.

The leeches are not very large, some 5–8 centimetres long, but they appear to be capable of speed, by stretch, tail-to-mouth, stretch, dozens of them and all heading for my legs, clearly smelling me a long distance away. I don't feel the leeches once they reach their destination (i.e. my legs), but they attach with a rubbery insistence and are hellishly difficult to remove, leaving a bloody reminder of their sojourn. Maddeningly, and because I have to watch my rather shy duck-billed platypus, I cannot stir, on pain of detection which would be the end of observations.

Clouds Creek is a remote, mostly slow-moving woody stream, somewhere deep inland on the plateau of New South Wales, in eastern Australia. There are large pools in the stream, their bottom covered in leaf litter from the huge gum trees, and around it there are wallabies hopping through the undergrowth. Every so often the odd wombat stomps through a clearing and large, black cockatoos, as well as the white sulphur-crested cockatoos, screech higher up in the canopy. But what keeps me here for days on end is none of these. It is just that agitating, floating blob in the water, with others of them further along.

A platypus is little more than half a metre long, including its duck-beak and its large flat tail. Water is its natural habitat, and what makes me think of it as a blob is its rotund body, of which part is visible above the surface when it is floating or paddling along. The tail is always underwater, and above the bill, which is munching continually, there are those two black, beady eyes, appearing rather aggressive and somewhat evil. For a brief time, the only movements are the concentric waves along the surface, emanating from the chewing bill.

I measure the animal's time on the surface with a stopwatch, and when it flips under into another dive, I move fast to escape the leeches. Within seconds I go down again behind a log, pressing my stopwatch when the platypus hits the surface once more, now further away.

My excuses for being here in Clouds Creek started a few months ago, in Armidale. There, I was studying the scent-marking behaviour of a curious marsupial carnivore, the spotted-tailed quoll, an animal about the size and much of the looks of a marten, living in the deep, immense gorges in the north-west of New South Wales. While I was looking around in the areas of farmland mixed into natural woodlands and riverbeds, I became intrigued by the often greenish water, and the unnaturally bright green colours along the waters' edges of willows and other trees. Nitrogen run-off from the farmland must be high, and the newspapers complain about the overuse of fertilizers. If the vegetation shows this, what about the animals living in these streams, and especially, what about that most uniquely Australian mammal, the platypus? Is an ugly wind from our civilization blowing its way?

I watch platypus in a few different places, in rivers and pools that are focal points for all sorts of life, with lots of the kind of rich tapestry that shows Australia at its best. My visits are in the early morning, with the waters steaming in the cold, upland air, and with the bell-like calls and screeches of many birds.

If platypus are affected by what agriculture does to these waters, then I expect that they will have feeding problems in the worst sites, so I decide to check up on that. Interestingly, the platypus's feeding is one of the weirdest among mammals. Many characteristics of the animal may look like backward steps in evolution (such as its egg-laying, lack of teats, and low body temperature), but the foraging behaviour of this incredibly efficient diving machine is fully twentieth century, using electrical fields.

Exploring the underwater branches, nooks and crannies, the platypus uses sense organs in the bill to pick up minute electrical fields around insect larvae, small crustaceans and molluscs. It is the only mammal in the world to use this technique. Having collected a bill-full, it surfaces and chomps its booty where I see it rippling the waters, then dives again.

Echidna, another animal unique to Australia. (Loeske Kruuk)

I reckon that in places where food is scarce, a platypus has to spend a longer time underwater for every mouthful. So when I watch, I time its dive and the following period on the surface: if the time underwater is relatively long and on the surface relatively short, then the platypus is likely to find its foraging more difficult.

That is where the platypus in the leech-infested Clouds Creek are so interesting. It is a marvellous bit of wild country, miles away from anywhere and little affected by the activities of man. Poor soils, in woods that are as wild and Australian as they come. Surprisingly and against my expectation, platypus have a more difficult time here in Clouds Creek than anywhere else I looked. But bless them – their long times underwater give me more opportunities to move about on the bank without disturbing them. For me it is a bit more time that enables me to avoid the ghastly leeches.

In other places nearer to civilization with its nasty nitrogen and other by-products, platypus have an easier life, with brief dives and long munching times on the surface. In the gorges near Armidale, a mile or so downstream of the outflow of the sewage works, is the splendid Blue Hole, a large pool with a beautiful atmosphere. Often there are wallabies hopping near the edge, an echidna plodding through the shrubbery and little cormorants spreading their wings over rocks in the water. For the many platypuses here, life could be called a doddle: comparing their diving periods in waters equally shallow as in Clouds Creek, it takes them about a third of the time underwater to collect a bill-full.

These two sites, Clouds Creek and Blue Hole, are extremes. Travelling around, I am having a tremendous time with the platypus in several other Australian places, which fall neatly in between the two in my measure of platypus foraging, and in their relative pollution (which was measured earlier by other scientists). I compare the life of this fabulous animal in the Blue Hole with platypus activities in small, spectacular and very food-poor lakes high up near Cradle Mountain in Tasmania. Similarly, I watch other platypus in forest streams at Eungella in Queensland. I come to realize that this species doesn't do too badly out of the by-products of civilization: in fact for the duck-billed platypus, eutrophic (i.e. food-rich) is best, even if artificial.

Of course, to me at least, a platypus may look much better, and is nicer to see, in a small lake, next to a pencil pine in the pure mountain air in Tasmania, or in a stream through a sweltering rainforest under giant logs and tree ferns, instead of in the rather polluted Blue Hole. But agriculture, or even the output of a sewage farm, does make platypus life somewhat easier. Perhaps, then, I do not need to worry too much on their behalf.

It is a different story for the spotted-tailed quoll, that strangely Australian, marsupial answer to our marten in Europe or America. It is the animal that draws me here (at least officially it is my research project). The platypus is just a very temporary diversion as an excuse for me to get away from the impossibly deep, rocky, hostile canyons in which the quoll keeps itself to itself. Climbing down from the plateau into the 200-metre-deep canyons is a hair-raising enterprise, though amply rewarded.

That the area where the quolls live is so almost impossibly difficult is no accident – it is the place where quolls can survive against the onslaught of all that mankind has thrown at them, such as agriculture, with foxes, cats, dogs and other exotics. At the bottom of the Salisbury River canyon at Dangers Falls near Armidale, huge rocks and boulders, crevices and cliffs provide safe havens. Once you have your eye in for these animals, you find their faeces on many latrines, on top of large flat rocks that are often bleached by urine, each of them with up to a dozen oily-smelling scats rubbed onto the stone and scattered.

With my friend Peter Jarman I clamber along from one latrine to the next. I am rather amazed at the similarity between these quoll latrines and what I have seen from totally unrelated animals on other continents, from hyenas, badgers or otters. Not only do these quolls look rather like the 'conventional'

Spotted-tailed quoll foraging. (Lyndon Meir)

carnivores in Europe or Africa (although as marsupials they are much more closely related to wallabies), they also behave like martens or other carnivores. At least that is what their latrines suggest. Conspicuous as these concentrations of scats are, they must have a scent-marking function. I am itching to be able to watch the quolls here in their wild places, but I realize of course that, alas, there is slim chance of following them in a place as inhospitable as this.

Looking at the contents of the scats left on the latrines, the carnivorous character of the quolls is obvious. Under the microscope in the laboratory I can identify hair of wallabies, possums and gliders, so quite large mammals are the mainstay of their diet. It seems highly unlikely that they would get all these things from scavenging, so they must be proper predators in their own right. And at least some – perhaps even most – of their prey must be caught on the areas around the deep gorge, or in the caverns, tree-holes and rock-clefts where the gliders and possums sleep.

We decide to put out a few, quite animal-friendly cage-traps near the latrines. This way, at least we can have a look at the quolls, and get some faeces from the animals caught to make sure that we have correctly identified the scats on the latrines. Baited with pieces of meat, we leave the traps out overnight next to some of the latrines, and heave ourselves back next day, climbing in and out of the horrendous gorge. And yes, we are in luck – the animals are remarkably easy to catch. They are a beautiful reddish colour, covered in large white spots.

Spotted-tailed quoll, an able climber.

What strikes me most of all is their tameness – they sit in the traps quietly, just waiting for us, and when I open the trap the quolls are very easy to handle. They are a world of difference from any of the small carnivores I am used to, from cats, martens, badgers or otters, which would have mauled my hands to shreds if I had picked them up like this.

That tameness, incidentally, I also notice when trapping Australian rodents in some of the forests to see what the potential prey fauna for quolls is like. Using large box-traps, which do not harm the animals in any way, I find it very striking that if I catch one of the introduced species, such as a brown rat, it is screaming, aggressive and noisy, attempting to attack and murder me. In contrast, a native Australian rodent just sits there quietly; I imagine a friendly grin on its face.

Perhaps this lack of overt aggression in the quolls and many other native mammals in Australia has something to do with the fact that also, there is not much evidence that they have territories. This is unlike carnivores and other mammals of similar size in Europe, Asia or America, which are highly territorial. The latrines of the quolls, if they do function as scent-marking stations, are clearly not markings of any territorial boundaries, and they are scattered all over the bottom of the gorge. Perhaps they are like the spraint sites of the Eurasian otters, which appear to inform other otters about who is feeding where, rather than acting as boundary markers. Australian researchers have shown that close relatives of the quolls, such as the Tasmanian devil, also don't have territories, and move about more or less randomly.

It is difficult to avoid the impression that these quolls and their other native compatriots just do not have the aggression needed to put up with the invading forces of non-native animals, of foxes and cats. These were introduced as a ghastly mistake by our forebears, and quolls also have to put up with a much earlier introduction, that of the dingo. The spotted-tailed quoll in Australia may be hanging on as a species by the skin of its teeth. In the face of depredations by exotics here in New South Wales, it is withdrawing into an almost impossible ravine habitat. I see quolls as the epitome of native Australian fauna in trouble – unlike the platypus, which may even do rather well out of what mankind does to the rivers.

The introduced carnivores are an unmitigated menace to Australia. Yet, despite these toxic additions, the country still holds a huge, beautiful treasure of wild nature. It is a treasure so different that it is almost unimaginable to a non-Australian.

CHAPTER 24

Just one last project

American mink.

THE BULKY FRAME of Nikolai darkens the low doorway of our small wooden house, against the snowy wilds outside. I am deep in Belarus, close to the border with Russia. Nikolai is a big man, though he is hardly more than a teenager. Over his shoulders he carries a large, dead animal, which he drops on the floor, unsmiling.

The lynx he brings in is the size of a small sheep. It is magnificent, with its beautifully pale, dense fur, its dark grey spots, and the long-tufted ears. Its eyes are glazed over. We stand around it, intensely sad. Nikolai trapped it, he said 'by mistake', and he wonders if we as zoologists would be interested; he himself just wants the skin.

More than anything else, it is this incident that makes me realize what a trivial little project I am involved in here in this vast Eastern European country.

My project is small in size and importance compared with the huge conservation problems in the country, or anywhere else in the world. I happen to be here to study European mink. But surrounded by seemingly endless, magnificent forests, inhabited by an array of mammals, of large predators, herbivores and multitudes of birds, all threatened by a poaching populace, the small European mink appears to be totally insignificant.

There are not many lynx left here, nor are there many bear, wolf or wolverine. They do still inhabit these wilderness areas, but desperately need protection. The forests bristle with cruel leg-hold traps and snares. It will be a battle of gigantic proportions to protect these animals from a people living in poverty, quite outside the range of anything I can do now. Somewhat lamely, I decide that small conservation problems also need solving and that, somehow, I myself have to concentrate, though it leaves me feeling faint and impotent. Later, making the best of a bad job, my Belarusian colleagues and I eat haunches of lynx (a delicate, veal- or pork-like flavour). We continue with our fieldwork.

The endless, flat Belarusian forest groans under a load of snow. Everything is dead quiet; there is no wind, no sounds, the sky is grey and heavy. Vadim Sidorovich moves his skis around fallen trees and thickets, following the course of a stream which is mostly covered by the white blanket, and I stay closely behind him. He has headphones over his ears, and he is waving a radio aerial. Today he has been going for three hours, trying without success to find a female European mink which carries a small radio-collar. He has been in contact with it almost daily for four months, usually along these same few kilometres of stream. Listening out for regular bleeps through the earphones he knows if the animal is within about a kilometre. At the same time both of us are keeping an eye out for tracks in the snow. So far today there has been no luck.

The quiet of the forest is deceptive, as Vadim's expert eye for snow tracks tells him. Our skis cross signs of roe deer, wild boar, elk, wolf, lynx, raccoon dog, fox, pine marten – a rich fauna. But no European mink. What we do find is a track that is different, though only slightly. It signals the new arrival of a close relative but different species, the American mink, this one a female. Hours later, we finally catch up with Vadim's European mink: she has moved miles away. The move is quite out of character and takes the animal a long distance from the stream, to a new site where it sleeps underneath a huge fallen oak tree. She is displaced. It happens every time when the American invader is about.

The European mink is a beautiful small, graceful and chocolate-brown animal with strikingly white lips. It is rare, very little known, and it is only to the eye of the initiated that it looks different from the introduced American mink, which we know so well in Britain. The American one is somewhat larger, stockier, often more black than brown, without white on the upper lips. In fact, as a species the European mink is related more closely to the polecat than to the American mink. Polecats look rather more different though, with longer fur; they are lighter in colour, and have a lovely masked face. All three of these marten species are much smaller than the otter, which is also common here.

During my time in Eastern Europe I never see the two mink species together but, according to Vadim's radio tracking, when they meet aggression between them is inevitable. It results in the European one on the run, the American invader as the winner. Polecats suffer little, and in any case they often live in somewhat different places, especially often around farms and other human dwellings.

The European mink is disappearing fast from almost all areas where it formerly occurred. Mostly, invasions of American mink are to blame, and I am part of a research project to find ways of reversing the trends. In Belarus I join Vadim, in Estonia I work with Tiit Maran who organizes the removal of all American mink from the island of Hiiumaa in the Baltic Sea, and the release there of his captive-bred European mink. Hiiumaa is large (1,000 square kilometres), quite far from the mainland, and the reason why it is overrun by American mink is that a local mink farm went bankrupt and opened its cages.

After an intensive trapping blitz catching 53 animals, the American mink is gone from Hiiumaa. In its place, European mink are now established. I fervently hope they will make it, and that the same can be tried elsewhere. There will be few further chances for the European mink, and action needs to come fast.

In Belarus, the endless forests still look as probably they did centuries ago. Wolves, bears and lynxes leave their tracks in the snow, elk browse the willows, cranes dance in the fields. Along streams and lakes there is an abundance of beavers, which with their large dams create prime habitat for mink and otter. There are otters, foxes, raccoon dogs and polecats everywhere – but all of them now hunted with increasing intensity, with any means available to the impoverished local people. I realize how much of an uphill struggle we have to fight to keep any of those treasures.

I love these forests, the remote wilderness. But for my science, for me to understand what is happening I also need to be able to watch the animals, to see their behaviour, to quietly sit and find out what they are doing and why. For me, that is the way in which I can establish what nature needs, so one can try to make sure it gets that. But I have a suspicion that if a species' survival is seriously threatened, it really is too late for me. Here in Belarus and Estonia, I find watching the behaviour of mink and other animals really difficult: local conditions make it almost impossible. Instead, I have to leave this mostly to other scientists who trap, and follow radio signals from a distance.

Release of a European mink on Hiiumaa.

Beaver dam in Belarus, creating prime habitat for mink and otter.

Apart from producing significant scientific results, personally I also want to do this in the field in a way that gives me an instinctive satisfaction. Like a journey, it is not just the getting there, the destination, it is the delight one experiences on the way. Some of my end results may be useful, but often afterwards it seems that in my projects I get most of the satisfaction from the doing rather than the publishing. Almost like travelling somewhere in a small plane flown by myself, when I can admire the action and the scenery wherever I may be going, my research needs to be full of background, of rich tapestry and of natural history.

That is where the Serengeti is so wonderful, or Ngorongoro, or Shetland or many other such sceneries. When watching animals there they dominate, I can feel that 'all this is theirs'. I hope that others, as well as my children and grand-children, will be able to experience this. I hope that some animal like Solomon will come their way.

Natural history brightens any human life, like music, and I took from my parents a passion for both. I am in the especially advantageous position of being able to use this enjoyment of nature in my research, and for that I have to thank Niko Tinbergen more than anyone. From him comes the questioning in the wilds of nature, questioning after watching, the quest for solutions to problems of behaviour and of survival. Put this in an African environment such as the Serengeti and there is the zenith of my life and passions.

Doing the research as work is wonderful; the world around me is magnificent. Some of the results of my probing may also be relevant, especially useful perhaps to other people wanting to walk the path of natural history and science. And also, I hope, to people who are actively involved in conservation, in whatever way.

References

Detailed references

Chapter 1
Tinbergen, N. (1951) *The Study of Instinct*. Oxford: Oxford University Press.

Chapter 2
Kruuk, H. (1963) Diurnal periodicity in the activity of the common sole, *Solea vulgaris* Quensel. *Netherlands Journal of Sea Research* 2: 1–28.

Chapter 3
Kruuk, H. (1964) *Predators and Anti-Predator Behaviour of the Black-Headed Gull*. Leiden: Brill.
——— (2003) *Niko's Nature: A Life of Niko Tinbergen and his Science of Animal Behaviour*. Oxford: Oxford University Press.
Tinbergen, N. (1951) *The Study of Instinct*. Oxford: Oxford University Press.
Tinbergen, N., Broekhuysen, G.J., Feekes, F., Houghton, J.C.W., Kruuk, H. and Sculz, E. (1965). Egg-shell removal by the black-headed gull, *Larus ridibundus* L., a behavioural component of camouflage. *Behaviour* 19: 74–117.

Chapter 4
Grzimek, B. and Grzimek, M. (1960) *Serengeti Shall Not Die*. London: Hamish Hamilton.
Kruuk, H. (1972) *The Spotted Hyena, a Study of Predation and Social Behaviour*. Chicago and London: University of Chicago Press (repr. 2014, Brattleboro, VT: Echo Point Books).
——— (1975) *Hyaena*. Oxford: Oxford University Press.

Chapters 6 and 7
Kruuk, H. (1966) Clan system and feeding habits of spotted hyaenas (*Crocuta crocuta* Erxleben). *Nature* 209:1257–8.
——— (1972) *The Spotted Hyena, a Study of Predation and Social Behaviour*. Chicago and London: University of Chicago Press (repr. 2014, Brattleboro, VT: Echo Point Books).
Lorenz, K. (1966) *On Aggression*. London: Methuen.

Chapter 8
Kruuk, H. (1964) *Predators and Anti-Predator Behaviour of the Black-Headed Gull*. Leiden: Brill.

———— (1972) *The Spotted Hyena, a Study of Predation and Social Behaviour*. Chicago and London: University of Chicago Press (repr. 2014, Brattleboro, VT: Echo Point Books).

———— (1972) Surplus killing by carnivores. *Journal of Zoology* 166: 233–44.

———— (2002) *Hunter and Hunted: Carnivore Behaviour and Their Relations with People*. Cambridge, UK: Cambridge University Press.

Chapter 10

Kruuk, H. (1976) Feeding and social behaviour of the striped hyaena (*Hyaena vulgaris*). *East African Wildlife Journal* 14: 91–111.

Kruuk, H. and Sands, W.A. (1972) The aardwolf (*Proteles cristatus* Sparrmann) as predator of termites. *East African Wildlife Journal* 10: 211–27.

Chapter 11

Kruuk, H. and Mills, M.G.L. (1983) Notes on food and foraging of the honey badger *Melivora capensis* in the Kalahari Gemsbok National Park. *Koedoe* 26: 153–7.

Mills, M.G.L. (1990) *Kalahari Hyaenas*. London: Unwin Hyman.

Chapter 12

Kruuk, H. (1980) *The Effects of Large Carnivores on Livestock and Animal Husbandry in Marsabit District, Kenya*. IPAL Report E-4. New York: United Nations Environmental Programme.

Chapter 13

Kruuk, H. (1995) *Wild Otters*. Oxford: Oxford University Press.

Chapter 14

Kruuk, H. (1967) Competition for food between vultures in East Africa. *Ardea* 55: 171–93.

Chapter 15

Grzimek, B. and Grzimek, M, (1960) *Serengeti Shall Not Die*. London: Hamish Hamilton.

Houston, D.C. (1974) The role of griffon vultures *Gyps africanus* and *Gyps ruppellii* as scavengers. *Journal of Zoology* 172: 35–46.

Pennycuick, C. and Pennycuick, S. (2016) *Birds Never Get Lost*. Leicester: Matador.

Van Lawick-Gooddall, J. and Van Lawick-Gooddall, H. (1966) Use of tools by Egyptian vulture, *Neophron percnopterus*. *Nature* 212: 1468–9.

Chapter 16

Kruuk, H. and Snell, H. (1981) Prey selection by feral dogs from a population of marine iguanas. *Journal of Applied Ecology* 18: 197–204.

Chapter 17

Kruuk, H. (1989) *The Social Badger*. Oxford: Oxford University Press.

Chapter 18
Kruuk, H. (2000) Note on status and foraging of the Pantot or Palawan stink-badger, *Mydaus marchei*. *IUCN Small Carnivore Conservation* 22: 11–12.
Kruuk, H. and de Kock, L. (1981) Food and habitat of badgers, *Meles meles*, on Monte Baldo, N. Italy. *Zeitschrift für Säugetierkunde* 46: 295–301.

Chapters 19 and 20
Kruuk, H. (1995) *Wild Otters*. Oxford: Oxford University Press.
——— (2006) *Otter Ecology, Behaviour and Conservation*. Oxford: Oxford University Press.

Chapter 21
Blundell, G.M., Ben-David, M. and Bowyer, R.T. (2002) Sociality in river otters: cooperative foraging or reproductive strategies? *Behavioural Ecology* 13: 134–41.
Kruuk, H. and Goudswaard, P.C. (1990) Effects of changes in fish populations in Lake Victoria on the food of otters (*Lutra maculicollis* Schinz and *Aonyx capensis* Lichtenstein). *African Journal of Ecology* 28: 322–9.

Chapter 22
Kruuk, H., Kanchanasaka, B., O'Sullivan, S. and Wanghongsa, S. (1994) Niche separation in three sympatric otters *Lutra perspicillata*, *L. lutra* and *Aonyx cinerea*, in Huay Kha Khaeng, Thailand. *Biological Conservation* 69, 115–20.

Chapter 23
Kruuk, H. (1993) The diving behaviour of the platypus (*Ornithorhynchus anatinus*) in waters with different trophic status. *Journal of Applied Ecology* 30: 592–8.
Kruuk, H. and Jarman, P.J. (1995) Latrine use by the spotted-tailed quoll (*Dasyurus maculatus*: Dasyuridae, Marsupialia) in its natural habitat. *Journal of Zoology* 236: 345–9.

Chapter 24
Maran, T., Macdonald, D.W., Kruuk, H., Sidorovich, V. and Rozhnov, V.V. (1998) The continuing decline of the European mink, *Mustela lutreola*: evidence for the intra-guild aggression hypothesis. *Symposia of the Zoological Society* 71: 297–323.

Index